IN SEARCH OF
VIKINGS

INTERDISCIPLINARY APPROACHES TO THE SCANDINAVIAN HERITAGE OF NORTH-WEST ENGLAND

Watercolour by W.G. Collingwood of the 'Thing-mound' at Fell Foot Farm, Little Langdale, Cumbria, north-west England. Courtesy of the Ruskin Museum (catalogue number ConRM 1989.807).

IN SEARCH OF VIKINGS

INTERDISCIPLINARY APPROACHES TO
THE SCANDINAVIAN HERITAGE OF
NORTH-WEST ENGLAND

EDITED BY
STEPHEN E. HARDING · DAVID GRIFFITHS
ELIZABETH ROYLES

CRC Press
Taylor & Francis Group
Boca Raton   London   New York

CRC Press is an imprint of the
Taylor & Francis Group, an **informa** business

Front cover: *Odin rides to hel.* Etching by W.G. Collingwood (Liverpool), 1908. Image made available on Wikimedia Commons by Haukur Þorgeirsson.

CRC Press
Taylor & Francis Group
6000 Broken Sound Parkway NW, Suite 300
Boca Raton, FL 33487-2742

© 2015 by Taylor & Francis Group, LLC
CRC Press is an imprint of Taylor & Francis Group, an Informa business

No claim to original U.S. Government works

Printed on acid-free paper
Version Date: 20141112

International Standard Book Number-13: 978-1-4822-0757-6 (Paperback)

Visit the Taylor & Francis Web site at
http://www.taylorandfrancis.com

and the CRC Press Web site at
http://www.crcpress.com

Contents

Foreword

This is a most impressive and very interesting book. *In Search of Vikings* shows how an interdisciplinary approach can provide a more nuanced understanding of the Vikings in peace and at war. To illustrate this, it focuses on one particularly exciting area or 'hot spot' of the Viking world, namely the north-west of England, an area where they settled in large numbers.

This fascinating book has been edited by three prominent experts with strong roots in this part of the country: Stephen Harding, Professor of Applied Biochemistry at the University of Nottingham and an expert on the scientific aspects of the Viking settlement period in this region; David Griffiths, Reader in Archaeology at the University of Oxford and one of the world's leading Viking archaeologists; and Elizabeth Royles, Keeper of Early History and Curator of Archaeology at the Grosvenor Museum in Chester. All three are from the Wirral Peninsula, an area steeped in Viking tradition following the initial settlement of Norwegian-led Vikings in the year AD 902.

This timely book brings together the various pieces of scientific, linguistic, historical, cultural, and archeological evidence to show how close Norway and the United Kingdom are. The strong relationship goes back hundreds of years and today we are close allies and trading partners, and the cultural links are rich and diverse.

I strongly recommend *In Search of Vikings,* which should be required reading for anyone interested in Viking history in the north-west of England.

Kim Traavik, Royal Norwegian Ambassador
London

Preface

The title of this book expresses the nature of Viking studies as viewed by its editors and contributors: it is a search, and not an easy one, and by no means a complete one as yet. A thousand years is a long time for the traces of a migrant Scandinavian people and culture to remain detectable. Most of the settlements, burials, and hoards the Vikings left behind gradually disappeared under more modern development or eroded away, leaving only a few startling discoveries to be made in modern times. Carved stones in churches have fared a little better, but many have still been lost to posterity. Language, folk traditions, and bloodlines have diffused with generations of later influences and new introductions, but some key aspects survive.

Putting all these types of evidence together is a major challenge. No individual is expert in every research skill and area of understanding required. It is, and should be, a team effort. Until fairly recently, archaeologists, place-name scholars, linguists, saga scholars, historians, and geneticists tended to work in their own separate disciplines, in some cases jealously guarding their territories. Few publications were brave or rash enough to attempt a genuinely interdisciplinary approach. We wanted to try to change that.

North-west England is the focus, but not the narrow obsession, of this book. We recognise that it is only a small part of the wider Viking world. Even in Britain, the north-west is perhaps overshadowed by York, the Danelaw, and Orkney and Shetland. Its near (and inter-visible) neighbour, the Isle of Man, gives its Viking past a higher cultural profile. Just across the Irish Sea, the great trading town of Dublin once stood pre-eminent in the western Viking world.

The Viking Age of north-west England has some intriguing facets: a number of distinctive localised settlement areas, with strong concentrations of Scandinavian place-names; some exciting excavations but many unanswered questions as to the extent of Viking settlement; rich surviving traces of Viking dialect in local speech and folklore; the largest Viking silver hoard yet known from Cuerdale (Lancashire); and a rash of recent hoard discoveries.

It has also produced some of the most promising biological data from genetic sampling programmes in the modern population, which is reason in itself to merit a reappraisal of the Viking influence in the region. Whatever flows from this will have much to say about north-west England and, we hope, potentially something useful to say about the rest of the Viking World.

The basis of this book was an interdisciplinary conference, held at the Grosvenor Museum, Chester, in November 2010, as a key part of the *Footsteps of the Vikings* project, funded by the UK Heritage Lottery Fund. The project aimed to highlight the Viking heritage of north-west England, share the most up-to-date research in the field with a wide audience, and forge links of institutions and individuals to promote the subject both during the project and as a legacy after it had come to an end. The conference titled 'In Search of the Vikings' achieved these aims by bringing together the most recent academic research focusing on the evidence for Viking presence and settlement in the region. It was open to all and was well attended by a varied audience who were able to enjoy the papers presented by leading experts in Anglo-Scandinavian history, genetics, landscape, and language.

The series of chapters here is largely based on lectures given at the conference. The purpose of the Chester conference was identical to that of this book: to bring new research on the Viking period in north-west England together for the first time, and make it accessible to a much wider audience than the researchers.

Sankthansaften (St. John's Eve) 2014
SH, DG, and ER

Contributors

Jo Buckberry
School of Life Sciences
University of Bradford
Bradford, UK

Paul Cavill
School of English and English-Place Name
    Society
University of Nottingham
Nottingham, UK

David Griffiths
Department for Continuing Education
University of Oxford
Oxford, UK

Stephen E. Harding
School of Biosciences
University of Nottingham
Sutton Bonington, UK

Judith Jesch
School of English
University of Nottingham
Nottingham, UK

Jane Kershaw
Institute of Archaeology
University College London
London, UK

Turi King
Department of Genetics and School of Historical
    Studies
University of Leicester
Leicester, UK

Christina Lee
School of English
University of Nottingham
Nottingham, UK

Ceilidh Lerwick
School of Life Sciences
University of Bradford
Bradford, UK

Mike McCarthy
School of Life Sciences
University of Bradford
Bradford, UK

Janet Montgomery
Department of Archaeology
University of Durham
Durham, UK

Caroline Paterson
Stirlingshire, UK

Robert A. Philpott
Museum of Liverpool
Liverpool UK

John Quanrud
School of English
University of Nottingham
Nottingham, UK

Roger White
Ironbridge International Institute for Cultural
    Heritage
University of Birmingham
Birmingham, UK

1

Interdisciplinary Approaches to the Scandinavian Heritage of North-West England

David Griffiths and Stephen E. Harding

CONTENTS

ABSTRACT   A wide variety of approaches based on historical, archaeological, place-name, linguistic, physical, and genetic tools can collectively provide a detailed understanding of the Vikings in peace and in war. This chapter and the rest of the book that follows focuses on one particularly exciting area of the Viking world, namely the north-west of England, an area where we now know them to have settled in large numbers. We take an overview of the technologies and approaches available and describe how new technologies are helping us to better understand what the Vikings left behind in terms of language, culture, archaeology, place-names, and genetic profiles of the people living there today.

Introduction

The Viking Age in England lasted for just three centuries, from the earliest recorded raids in the AD 790s to the Norman take-over that took place between the Battle of Hastings in 1066 and William Rufus's annexation of Cumbria in 1092. This 300-year period began with sporadic Viking raids, and saw increased political and economic interaction, settlement, and ultimately assimilation. Seen against the long-term background of two millennia of Roman, medieval, and modern history, the Viking Age was a relatively transient phase occurring when the national institutions and allegiances that we know today were barely beginning to take shape; it left a highly fragmentary and tantalizing set of physical and cultural traces.

The prominent and highly distinctive historical reputation of the Vikings, as it exists today, is out of all proportion to the muted and often contradictory evidence for their actual presence that has survived in historical, archaeological, and biological forms. An outstanding example of this—and the geographic focus of this book—is the north-west of England (Figure 1.1).

Attempts to quantify the Viking impact in England have encountered the difficulty of identifying historical Scandinavian influence clearly against an all-too similar Anglo-Saxon background. The Anglo-Saxons, of course, owed much of their cultural, linguistic, and biological inheritance to earlier waves of immigration and cultural influence from the North Germanic world including Scandinavia. A fierce

FIGURE 1.1 North-west England: Viking Age sites and political territories in their regional context. (Courtesy of Michael Athanson, University of Oxford.)

debate continues as to the extent to which incoming Anglo-Saxon peoples and cultures assimilated the native and Romanized Britons who already inhabited Britain.

The challenge of detecting the Viking presence against such a complex existing picture is further complicated by the fact that Vikings themselves did not necessarily all conform to a single biological or cultural stereotype. The word *Viking* comes from the Norse word *Vik* (bay or inlet) and came to refer to a seaborne adventurer or marauder (Brink 2008). To be a Viking, therefore, was more like a reputation or an occupation rather than an innate biological status.

Indeed, few if any of the Scandinavians active in Britain would even have identified themselves primarily as Vikings. Kin and regional affinities were more important than a vague catch-all term that almost certainly has far more meaning today than it had a millennium ago. In the political and religious melée of contemporary Britain and Ireland, with hostages, slaves, and mercenaries abounding, it was therefore possible to cross over and adopt Viking culture by choice or by compulsion without ever having been to Scandinavia.

We now understand that Vikings were not exponents of a monolithic, alien culture, but constituted a multiplicity of small groups that interacted with Irish and British societies on a varied and circumstantial basis, often showing considerable aptitude for intermarriage and rapid assimilation. Yet Scandinavian-derived traits persisted, often echoing pagan stories and motifs in art, inscriptions, and material culture.

Hybridity between the Scandinavian diaspora and native cultures is a complex and fascinating area of study across the Viking world from Greenland to Russia, and one that is now being re-shaped by an increasing flow of genetic and isotopic data. This book seeks to highlight and discuss a number of interdisciplinary challenges in bringing together archaeological, biological, historical, and linguistic evidence and emphasise that a nuanced picture, taking account of regional and historical diversity, is essential to understanding this period.

Perceptions of Viking Heritage

Across northern Britain, there are areas, cities, and communities that uphold a particular pride in having Viking heritage. Prominent amongst these are York, the Isle of Man, Orkney, and Shetland. In these places, Vikings are seen as local folk heroes and serve as a basis of a considerable tourist industry. Museum displays, heritage trails, and school history projects emphasise Viking themes. Festivals, perhaps most vividly exemplified by the annual winter ship-burning ceremony of *Up Helly Aa* in Shetland, give colour and drama, and reinforce the place of Vikings at the heart of contemporary local culture, distinctiveness, and communal self-esteem.

Notwithstanding the role of the Victorians and more recent civic enthusiasts in embellishing the facts underlying the growth of popular appreciation, there is a real historical resonance to these re-awoken traditions that enthuse and delight locals and visitors year upon year. Even in areas that make less of a regular habit of festive Viking commemoration, hints of Scandinavian heritage are widely (if imperfectly) understood and appreciated.

Norse place-names and church dedications such as Chester's St Olave (Olav or Olaf) inspire adults and school children to spend time reading about and creating art works alluding to the long-ago Viking presence. Since 2008, every July 29th—St. Olav's Day—or thereabouts, hordes of enthusiasts have trudged the 20 or so miles between West Kirby, Wirral, via Thurstaston and Neston to St. Olave's Church in Chester. Wirral's own version of Norway's St Olav Pilgrimage from Oslo to Trondheim (approximately 400 miles) was captured in Michael Wood's 2011 BBC series *The Great British Story* (Figure 1.2).

Popular culture has taken the Vikings to heart. Supporters of Football and Rugby League clubs such as Tranmere Rovers and Widnes Vikings wear plastic horned helmets at games and roar their support in a suitably war-like manner, especially when the match is against a club from an area lacking such an ebullient northern heritage. In 2002, a local brewery in Liverpool distributed commemorative beer mats to mark the 1100th anniversary of the arrival of the Vikings—with instructions as to what to do if the Vikings return (Figure 1.3).

The popular name for Liverpudlians, Scousers, is widely held to be of Viking origin, although in actuality this belief is probably misplaced. The potato stew called *Lapskaus* that gave rise to the term is more likely to have been introduced by Scandinavian sailors in more recent centuries (Harding and Vaagan, 2011). Perhaps the most remarkable indication of popular enthusiasm for Vikings has been seen in the responses of over 200 present-day residents of selected areas of north-west England to a call to provide DNA samples (from cheek swabs) as a means of researching the extent to which Scandinavian genetic traits occur in the modern population.

Fifty or more years ago, few places that now identify themselves strongly with a Viking past were noticeably doing so. Scandinavian place-names, archaeological evidence for the Viking presence, and linguistic survivals in speech and dialect, of course, existed as much as they do today (in the latter case even more richly then). It has taken the spread of the mass media, in part responding to the greatly increased profile of archaeological excavation from the 1960s onwards, plus a huge outpouring of academic and popular books and museum exhibitions, to reawaken interest in the Vikings. This has perhaps been assisted by a greater secularisation in modern society which has become less averse to glorying safely in the delights of paganism, epic slaughter, and fantasy heroism, than would more churchgoing-minded former generations with direct experience of devastating conflicts.

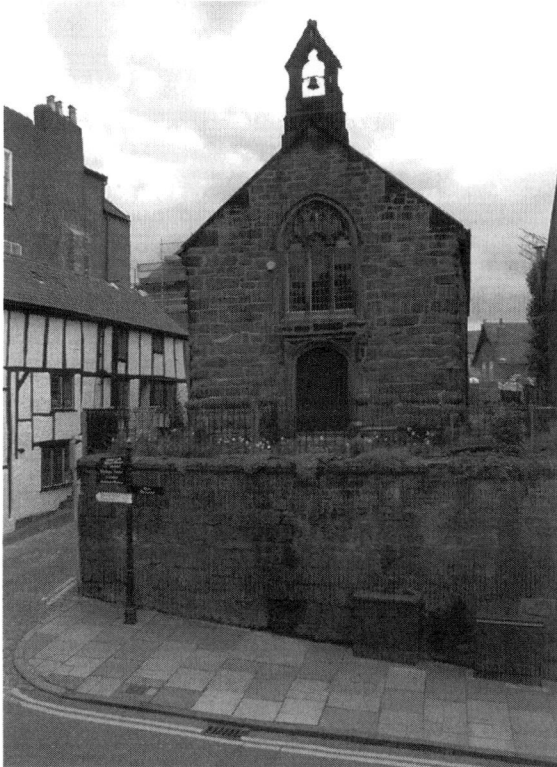

FIGURE 1.2   This time they come in peace! Top: St. Olav's Day walkers enter Chester on 29 July 2011 at the end of a 13-mile walk from Neston, Wirral to commemorate Norway's patron saint and help preserve the continued existence of a St. Olave Church in the north-west of England. They are greeted by broadcaster and historian Michael Wood. (Photo courtesy of Dan Kemp.) Bottom: St. Olave's Church, Chester.

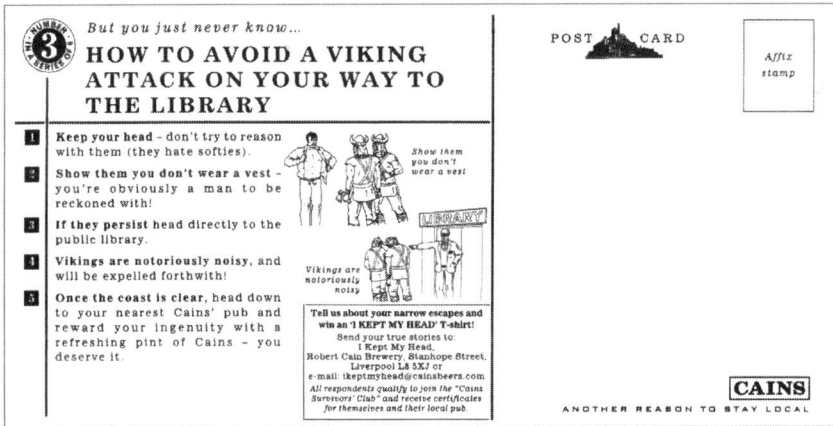

FIGURE 1.3 Beer mat produced by a local brewery in Liverpool to commemorate 1100 years after the arrival of the Vikings. The reverse side indicates what to do if they come back again. (Courtesy of Cain's Brewery, Liverpool.)

Gradually, awareness of a sharing in a Viking past has spread beyond its most visible and well-known centres in Britain and Ireland. This has not always been a smooth or trouble-free process. In Dublin in the late 1970s, considerable controversy and even some violence surrounded the city corporation's plans to redevelop the run-down Wood Quay area of the urban waterfront. Excavations from the late 1960s onward had shown that this area represented a deeply stratified treasure store of archaeology, with astonishingly well-preserved evidence for streets, houses, and quaysides dating back over 1,000 years to the time when Dublin was a Viking kingdom and the largest international trading port in the western seas (Bradley 1984; Johnson 2004). The plan to impose a modernist concrete municipal office development over this area attracted widespread protests and a sit-in prevented progress for a while. Confrontations between protestors and developers resulted in damage to property and even to some (fortunately not too serious) bloodshed. A less-than satisfactory compromise was eventually reached. It allowed the National Museum's archaeologists more time to excavate the site, but in some ways the lasting impact of these dramatic events was on popular perceptions of Irish history.

Led by a clergyman, Fr. F.X. Martin, the pro-Viking Dublin campaign (perhaps ironically) undermined the unquestioning historical assumptions behind the notion of 'The Land of Saints and Scholars.' A national origin myth of Celtic Christian purity underlying the culturally nationalist and politically republican politics of the previous century in Ireland was challenged seriously on a popular level for the

FIGURE 1.4 Thingwall, Wirral. Following popular support and financed by a local company, four special signposts were erected in May 2012. This one at Cross Hill marks the site of one of the leading candidates to be the site of Wirral's Thing. (Photo courtesy of Brian MacDonald.)

first time. Vikings (formerly viewed with disdain as barbarian outsiders and forerunners of Ireland's 750-year foreign occupation and colonisation by Normans, English, and Scots), were reclaimed as prestigious and dynamic forebears by ordinary Dubliners; they became objects of intense interest and have remained so ever since.

In north-west England, the principal geographic subject of this collection of papers, no such dramatic political confrontations have propelled Vikings suddenly into modern consciousness. Nevertheless, from Cheshire to Cumbria, there is a growing appreciation of the value and interest of the distant Viking past and its legacy today. From Thingwall to Ormskirk to Kirkby Lonsdale, people recognise and appreciate that they live in places that owe their names and historic identities to the Viking presence, and this is reflected in posting of heritage signs such as four recently erected on Wirral denoting Thingwall as the 'Assembly Field' (Figure 1.4).

Words in regional and local rural spoken dialects still preserve ancient connections and represent living links to the distant past. In 2001, BBC presenter Julian Richards, in filming the *Blood of the Vikings* programme, encountered in Cumbrian villages phrases probably more understandable to Norwegians than to modern Londoners, such as: 'Did thee come up here lakin when thee was a barn?' (Did you come up here playing when you were a child?).

Across the north-west of England there is local pride in the presence of Viking-Age sculptured stones in parish churches (many of which have recently been brought out of dusty obscurity with better lighting and interpretation panels) and a rush of interest every time an archaeologist or metal detectorist finds something that might reveal the Viking presence. Museum archaeologists and the regional finds liaison officers of the Portable Antiquities Scheme receive more enquiries from members of the public claiming to have found Viking objects than enquiries about almost any other period. Sometimes Viking attributions are correct, but very often the objects turn out to be misplaced from another archaeological period or are modern, but the reports continue to come in nevertheless.

The Wirral and West Lancashire Project for sampling Y-chromosomal DNA in the modern male population (Figure 1.5) took place from 2002–2007.* The results were published in 2008 (Bowden et al. 2008) and in popular form two years later (Harding, Jobling and King 2010). Its connection with the local archaeology of the region was also considered in a specially commissioned article for *British Archaeology* (Griffiths, Harding and Jobling 2008).

Over 200 men from old Wirral or West Lancashire families took part. All had paternal grandfathers from one of these regions. The group included over 80 individuals who possessed surnames that were present in these regions prior to 1600—names such as Totty, Forshaw, Sherlock, Robinson, Raby, Melling, Scarisbrick, Crombleholme, and Altcar. The pre-1600 surname factor was used to help circumvent the large population movements that occurred in north-west England since the Industrial Revolution

---

* www.nottingham.ac.uk/-sczsteve/survey.htm

(a)

(b)

FIGURE 1.5 DNA testing. (a) West Kirby: Volunteer from Wirral providing a cheek swab for the Wirral survey. (b) Ormskirk: Volunteers from West Lancashire completing appropriate documentation before providing their samples. DNA sampling must follow strict ethical guidelines.

and the tremendous growth of the Liverpool conurbation. By linking genes with modern geography in this way, could characteristic genetic signatures for the Norse settlers from a millennium previously be found? And to what extent? Turi King (Chapter 11) takes another look at the findings and explains the technologies involved.

Ground-breaking advances in population genetics for assessing genetic ancestry and links between populations (Jobling et al. 2013) have attracted considerable interest and debate. Sceptics have also wondered about the time when such traits may have appeared in the population. Researchers have had to be on their guard against attempts to twist their data for racially or culturally divisive ends, although much of the DNA (mitochondrial DNA or from male Y-chromosomes)-tested bears no relation to how people look or behave.

However, the idea that Viking blood and genes are still alive within current generations has a powerful allure to many, and this appeal runs wider than just those whose cheek swab samples revealed signature Scandinavian genetic traits. Others too, including children and adults of biological Viking heritage in other parts of the world, have enthusiastically adopted Vikings as mascots for their communities, sports

FIGURE 1.6   North-west England's modern Viking Navy rowing the *Draken Harald Hårfagre*, the largest contemporary Scandinavian longship (35 m long, 8 m wide) at the Karmøy Viking Festival, 2013.

clubs, and schools. North-west England even boasts its own modern 100-strong 'Viking Navy' of local enthusiasts expertly trained by the Liverpool Victoria Rowing Club on the River Mersey. In June 2013, this formidable group 'raided' Karmøy in western Norway, taking the opportunity to row the largest working Scandinavian longship reconstruction (Figure 1.6).

How, therefore, did such a strong contemporary popular perception of a Viking heritage come about, and to what extent do popular perceptions reflect the reality of a modest, fragmentary, and often frustratingly unclear stock of actual evidence for the Viking presence a millennium ago?

Searching for Vikings in North-West England:
The Development of Scholarly Enquiry

Between the Middle Ages and the dawn of modern scholarship, Vikings almost disappeared entirely into the background. Glimpses of their historical reputation and prestige survived in folklore and lingered on in the writings of a tiny number of learned scholars.

William Camden made extensive mention of the Danes in Britain in his *Britannia* of 1586. There was therefore still an awareness of the historical presence of Danes and 'Northmen' amongst antiquarians in the 17th and 18th centuries, when discoveries such as the Harkirke silver hoard from Little Crosby (Lancashire, now Merseyside), and the Viking burial mound at Aspatria (Cumberland, now Cumbria), were unearthed (Griffiths, Chapter 2, this volume).

It was, however, the romantic and artistic sensibility of the later 19th century that brought about a fuller resurgence in interest in the history, literature, and language of the Vikings in north-west England (Wawn 2000). The vicar of Aspatria (Cumbria), Reverend William Slater Calverley (1847–1898), brought the Viking Age stone sculpture of Cumberland and Westmorland into public and academic consciousness with a series of papers and observations, and by carving a much-admired sandstone replica of the Gosforth Cross that still stands in Aspatria churchyard. A near-contemporary of Calverley's and an even more influential and well-connected Old Norse scholar of Liverpudlian origins was William Gershom Collingwood (1854–1932). Collingwood (Figure 1.7), a devoted follower of John Ruskin and a luminary of the Arts and Crafts Movement, was a polymath who resided in the Lake District for much of his life (Townend 2009). An artist, folklorist, historian and field archaeologist, he arguably did more to rescue the Vikings of north-west England from obscurity than anyone else. Collingwood's exquisite drawings (Figure 1.8) and observations of Viking sculptured crosses and hogback grave markers culminated in his work of 1927: *Northumbrian Crosses of the Pre-Norman Age*, with a follow-up piece on Wirral's most extensive Viking sculpture group titled *The Early Monuments of West Kirby* (1928).

Collingwood's field research at upland sites such as Ewe Close and Crosby Ravensworth (Westmorland, now Cumbria) remain important contributions today. His travels in search of Vikings also took him to Iceland in 1897, where he explored the sites where the great sagas were written—painting and drawing many of them (Figure 1.8) He also painted an altar piece in pre-Raphaelite style that remains *in situ* at Borg Church, Borgarnes, the home farm of Egil Skallagrimsson of *Egil's Saga* (Cavill, Chapter 6). Collingwood's work is reflected in the *Victoria County History* of Cumberland and Lancashire published between 1901 and 1914 (these volumes were joined nearly a century later by the five Cheshire volumes of 1979–2005. New work on Westmorland is underway at the Centre for North West Regional Studies at Lancaster University).

Perhaps the most significant contribution to studies of the Viking presence in north-west England, which remains profoundly influential today, is the work of historian and place-name scholar Frederick Threlfall Wainwright (1917–1961). A South Lancastrian from Rainford, Wainwright (Figure 1.9) studied at Reading University in the late 1930s under Sir Frank Stenton, the great historian of Anglo-Saxon England and co-author of the EPNS* Cumberland place-name volumes (below).

Wainwright subsequently taught English in a Liverpool school during the Second World War, before going on to lecture at Dundee and St Andrews. He took up studies of Pictish Scotland in his later years before his untimely death, but remained principally throughout his life a historian of Lancashire and Cheshire. His papers titled 'Wirral Field Names' (1943), 'The Scandinavians in West Lancashire' (1946), and 'Ingimund's Invasion' (1948) drew renewed attention to the nearly forgotten history of the Vikings in Wirral and South Lancashire.

His rediscovery of the importance of O'Donovan's 1860 translation of the *Three Fragments of Irish Annals* (Griffiths, Chapter 2) remains a seminal contribution. Several of Wainwright's published and unpublished manuscripts, including 'The Field Names of Amounderness Hundred' were collected and

---

* EPNS: English Place-Name Society.

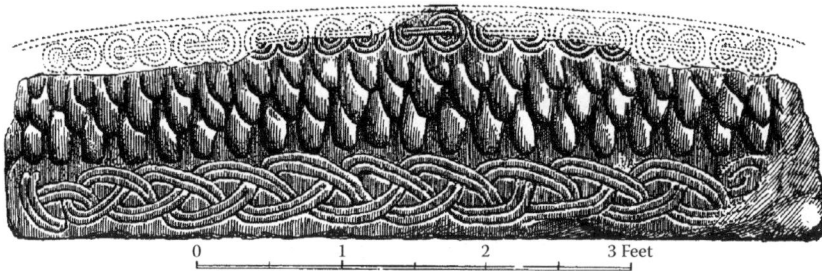

FIGURE 1.7   Top: W.G. Collingwood, antiquarian, born Liverpool 1854, died and was buried at Hawkshead, Cumbria, 1932. Portrait (c. 1881–1885) made by J.A.P. Severn. Courtesy, Ruskin Museum, Coniston, Cumbia. Bottom: Collingwood's sketch of the hogback tombstone at West Kirby.

edited by H.P.R. Finberg in 1975, providing a lasting memorial and perhaps filling the void of the important book on the Vikings in north-west England which Wainwright may have written had he lived longer.

Wainwright's rescue from obscurity of the Ingimund legend (Griffiths, Chapter 2), the subject of his 1948 *English Historical Review* paper, is probably his most memorable achievement. Wainwright's work, together with those of other historical and place-name pioneers mentioned above, provided a basis for later integrated historical and toponymic studies such as those by Gillian Fellows-Jensen (1985), Denise Kenyon (1991), Mary C. Higham (1995), Nick Higham (1993, 2004a, 2004b), and Angus Winchester (1987).

Wainwright's meticulous research on the place-names of north-west England was not their first published coverage, but built upon existing compendia to an unprecedented level of detail. A milestone in onomastic research was achieved earlier in the 20th century by Eilert Ekwall (1877–1964), a professor of English at Lund University in Sweden. His volume on the place-names of Lancashire, published by Manchester University Press in 1922, followed a seminal 1918 paper published in Lund, titled 'Scandinavians and Celts in north-west England.' Ekwall's Lancashire volume remains a monument to toponymic research on north-west England.

It was joined later in the 20th century by volumes produced by the English Place-Name Society, all based on the pre-1974 county boundaries (the post-1974 counties of Merseyside and Greater Manchester were split between Cheshire and Lancashire). Cumberland was covered in two volumes in 1950–1952 by

(a)

(b)

FIGURE 1.8 Some of Collingwood's drawings and paintings of the saga-steads of Iceland. (a) Drawing of Hlida-Endi from Gunnars-holm. (b) Hval-fjord from Brekka. (c) Painting of Borg, Iceland, home of Egil Skallagrimson, Iceland's most famous Viking. (d) Painting of the Thing at Thingvellir. These names are also reflected in place names of north-west England. *(continued)*

A.H. Armstrong, A. Mawer, Sir Frank Stenton, and Bruce Dickins. A.H. Smith, a Reader in Scandinavian studies at London University, wrote about Westmorland for a volume published in 1967. Cheshire was the subject of five volumes published between 1970 and 1981 by another London academic, John McNeal Dodgson (Figure 1.10), who earlier (in 1957) published an influential paper on 'The Background to Brunanburh,' arguing eloquently for Bromborough, Wirral as the battle site.

Dodgsons's theory was bolstered by an article titled 'Revisiting Dingesmere' (Cavill, Harding and Jesch 2004). Paul Cavill considers the issues afresh in Chapter 6. The reference in the *Anglo-Saxon Chronicle*'s account of the battle to 'Things Mere' is also a reminder of the Thingvatn (Þingvatn) of Iceland that may be seen in Collingwood's beautiful painting of Thingvellir (Figure 1.8).

The stone sculptures of north-west England, which were illuminated and illustrated by Calverley and Collingwood in the 19th century, also received attention in the mid to later twentieth century, not least by John D. Bu'Lock, a Reader in Chemistry at the University of Manchester (Figure 1.11) who summarised the Cheshire school of red sandstone sculptured crosses in 1958 and also wrote an important article on the north Wirral coastal site of Meols (Bu'Lock 1958 and 1960; summarised in Bu'Lock 1972).

(c)

(d)

FIGURE 1.8 (continued) Some of Collingwood's drawings and paintings of the saga-steads of Iceland. (a) Drawing of Hlida-Endi from Gunnars-holm. (b) Hval-fjord from Brekka. (c) Painting of Borg, Iceland, home of Egil Skallagrimson, Iceland's most famous Viking. (d) Painting of Thing at Thingvellir. These names are also reflected in place names of northwest England.

New discoveries of stone structures at Neston and Bidston were made by, respectively White (1986; Chapter 12, this volume), and Bailey and Whalley (2004). All previous sculptural studies were subsumed into the British Academy's *Corpus of Anglo-Saxon Stone Sculpture,* which provided the first comprehensive study of Viking Age stone monuments since Collingwood's work. The *Corpus* exists in two volumes for north-west England, separated by a gap of 22 years, on Cumberland, Westmorland, and Lancashire north of the Sands (Bailey and Cramp 1988) and Cheshire and Lancashire (Bailey 2010).[*]

---

[*] Using pre-1974 county boundaries.

FIGURE 1.9   F.T. Wainwright (1917–1961) on an excavation at Dundarg, Aberdeenshire, Scotland, 1951. (Copyright RCAHMS).

FIGURE 1.10   John McNeil Dodgson (1928–1990).

FIGURE 1.11   John Desmond Bu'Lock (1928–1996).

Before the late 20th century there had been almost no archaeological excavations which sought directly to address the Viking presence in north-west England, leaving no parallel to studies of stone sculpture or historical and onomastic research. The excavations of Gerhard Bersu on the 'Three Viking Graves' on the Isle of Man at the end of the Second World War created renewed interest in the Viking Age of the Irish Sea coastlands, although these excavations remained unpublished for two further decades (Bersu and Wilson 1966).

It took the urban redevelopments of the 1960s onward to begin to reveal more evidence in north-west England, with significant excavations taking place in Chester, particularly in the 1960s and 1970s (Mason 1985; Ward 1994), and Carlisle in the 1980s (McCarthy et al., Chapter 9). Other excavations, such as at the early medieval ecclesiastical site at Heysham, Lancashire, in 1977 and 1978 (Potter and Andrews 1994), at Workington, Cumbria, in 1996 and 1997 (McCarthy and Paterson, Chapter 8), and at Cumwhitton, Cumbria (Patterson et al. 2014) added to the picture.

Interdisciplinary Approaches in Modern Research

In the last few years, the volume of published literature on the Viking Age in north-west England has increased markedly, with several recent syntheses covering this area (all or in part) as case studies in their own right, as part of the Irish Sea region (for example, Graham-Campbell 1992; Cavill, Harding and Jesch 2000; Graham-Campbell and Philpott 2009; Griffiths 2010), or as part of the wider English picture (Richards 1991; Hadley 2006).

Arising from these studies is a sure and agreed consensus that if we are to make further progress in this field of research, we must approach it as a fundamentally interdisciplinary challenge. Historians, linguists, archaeologists, physicists, chemists, and biologists have begun to come together in a common quest.

With this increase in both the numbers and capabilities of approaches, several researchers have stressed the need for experts in particular disciplines to develop a familiarity of other methods outside their own fields of expertise to develop cohesive, interdisciplinary investigations with successful, well-understood communications. With this in mind, we now provide a brief outline of the spectrum of tools available to researchers of the Viking Age and indicate how an interdisciplinary integrated approach is being successfully applied to learn about the Viking presence.

History, Language, and Names

Vikings who raided and settled in north-west England from the 8th to the 11th centuries left no written records themselves. While large numbers of runic inscriptions are seen on the Isle of Man and in Dublin, they rarely exceed more than a few words. In north-west England, apart from one small inscription (a *futhark* or runic alphabet) on a silver brooch from Flusco Pike, Newbiggin, Cumbria, these consist of fragmentary inscriptions from a handful of Cumbrian churches, all of which are relatively late, from the 12th century, and tell us little about the earlier Viking presence (Barnes and Page 2006).

The earliest developed accounts of the period written in the Norse language are the Icelandic sagas that appeared 200 to 300 years later. The sagas purport to describe events in the Viking Age proper, and include, for instance, the story of Egil's participation in the Battle of *Vinheithr,* which was almost certainly the battle we know as *Brunanburh* (Cavill, Chapter 6). The sagas were based on an awareness of earlier events but were written to please an audience in their own time, glorifying the past roles of certain wealthy families important in the 13th century. Most scholars are cautious about fully accepting their embellished and semi-fictionalised accounts of history.

The main contemporary or near-contemporary sources of information on Vikings in north-west England are found in Irish, Anglo-Saxon, and Welsh annalistic sources, which sometimes mention the same events and therefore corroborate each other. The annalists were supportive of their own rulers and patrons and therefore tended to produce oppositional and highly coloured accounts of Viking activity. These are complex sources which tell and re-tell many versions of events. Charters, land grants, and wills of Anglo-Saxon authorship occasionally obliquely allude to the presence of heathen outsiders. To the latter we may add histories of varying reliability composed after the Norman Conquest such as those of William of Malmesbury, Simeon of Durham, and Florence [John] of Worcester.

Therefore, everything we know historically about north-west England in the Viking Age comes from a collection of written records created elsewhere, mostly some time after the events they describe. This inevitably led to some diffusion of detail and potential confusion of orders and locations of events. Far from giving us a consistent record, these records at best offer glimpses of real events separated by some-times entire decades of silence.

Major events such as ferocious battles involving the topmost ranks of rulers on both sides tend to figure more prominently than the more humdrum events in a day-to-day story. The gaps in the historical record leave us without clear answers to important questions. Were there Viking settlers in north-west England prior to the early 10th century? How united were the Viking settlers amongst themselves? How many Scandinavians settled in this part of England?

As demonstrated by Quanrud and Cavill's contributions (Quanrud, Chapter 5; Cavill, Chapter 6), it is possible to make convincing progress in building a historical picture by combining contemporary sources with later ones and carefully testing their convergences. Histories and poetry compiled at a somewhat later date along with occasional glimpses in saga literature, of course, may owe much to ear-lier and contemporary (but now disappeared) sources on which their authors base their accounts.

In north-west England, one remarkable source of evidence pointing at survival of a large number of dialect words of Norse origin appears to be the 14th century poem, *Sir Gawain and the Green Knight.* The general setting of the poem is across the northern Anglo-Welsh borderlands. Although the identity of the poet is unknown, a study of the language and dialect of the poem has led many experts to believe that he or she may have come from somewhere in the region of southern Lancashire and Cheshire.

Some have linked the authorship of the story with the Knight of the Garter, Sir John Stanley (1345–1413) of Storeton Hall, Wirral (see, for example, Wilson 1979). Stanley was a knight who served both Richard II and Henry III, and part of the action of the poem takes place in Wirral. The incorpora-tion of a large number of Norse dialect words (Harding 2002, pp. 182–184; Harding and Vaagan 2011; Dance 2013) makes the language of the poem tangibly distinct from Chaucer's *Canterbury Tales* written at around the same time.

As with the order and chronology of historical events, place-names also remain subject to important caveats about their dates and uses to denote Viking presence (Abrams and Parsons 2004). Place-names may be divided into topographic names and habitative names denoting settlement. Their distribution across England is strongly regionalised (Figure 1.12). Apart from those listed in *Domesday Book* of 1086, most settlement names were not recorded historically until the high Middle Ages (14th and 15th centuries) or later.

Common topographic names such as *fell, beck, dale, scar,* and *skerry* reflect the entry of Norse terms into area dialects or those of specialist occupations such as upland farming or seafaring, but lacking the

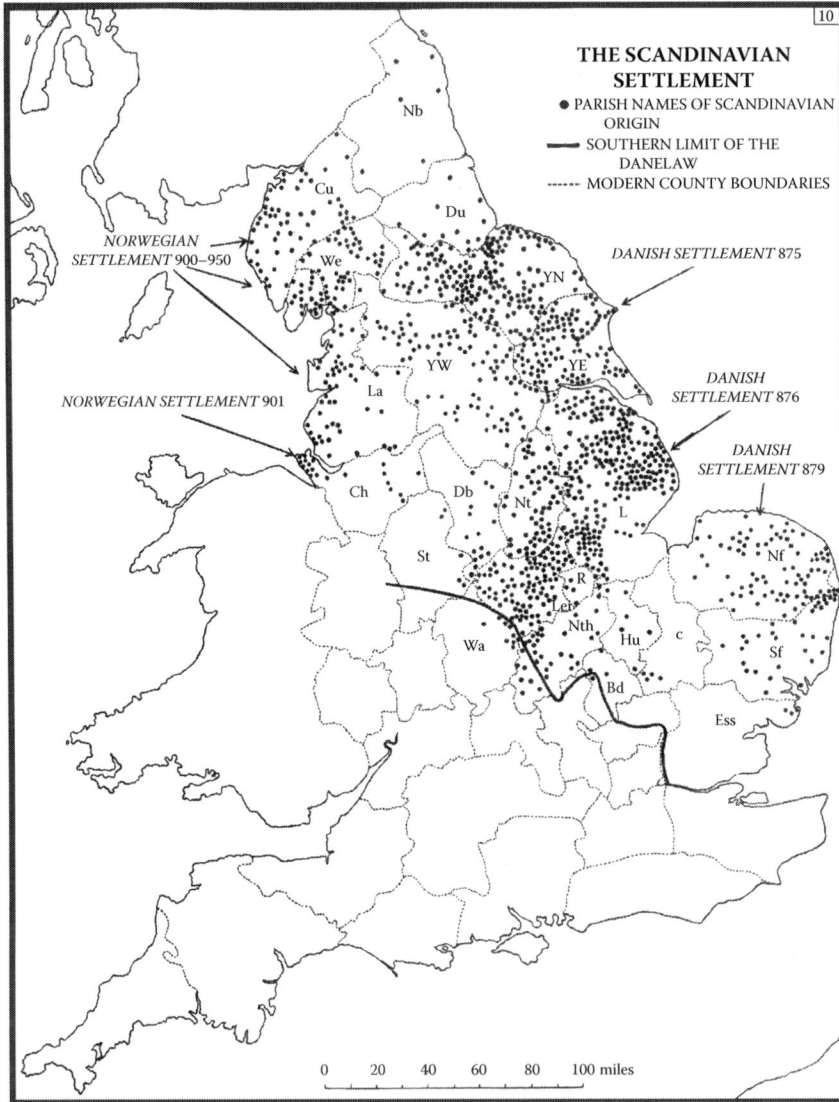

FIGURE 1.12   A.H Smith's 1956 map of Scandinavian settlement of England (reproduced by permission of the English Place-Name Society). Bd = Bedfordshire. C = Cambridgeshire. Ch = Cheshire. Cu = Cumberland. Db = Derbyshire. Du = Durham. Ess = Essex. Hu = Huntingdonshire. L = Lincolnshire. La = Lancashire. Lei = Leicestershire. Nb = Northumberland. Nf = Norfolk. Nt = Nottinghamshire. Nth = Northamptonshire. R = Rutland. Sf = Suffolk. St = Staffordshire. Wa = Warwickshire. We = Westmorland. YE = East Yorkshire. YN = North Yorkshire. YW = West Yorkshire.

legal status of landholdings, they tend to be cited in documents much later and are therefore harder to pin down to a specific date of origin.

However, patterns of density, even of very minor topographic names, can help identify some areas where Scandinavian speakers were more predominant in the past. Local names for topographic features such as *carr, rake*, and *holm* (Harding 2007) are common enough to imply the historical influence of Old Norse extending far beyond just a small, if powerful, Viking elite. The Scandinavian place-names of Cumbria, Lancashire, and Cheshire became fixed and accepted within the medieval English dialects spoken in these areas.

Place-names ending in *-bý*, such as are found in profusion in Wirral (Figure 1.13) and the West Derby hundred in south-west Lancashire have traditionally been seen as markers of Danish or Danelaw

FIGURE 1.13   Signpost at Irby, Wirral. All the names including Irby are Norse or Norse influenced.

influence. However the subtly different West Norse *-bær* and East Norse *-býr*, which are also rendered in modern times as *-by*, complicate the picture.

To confuse it still further, plantations of people from eastern England after the Norman Conquest imported the nomenclature of the Danelaw, notably around Carlisle and in the Eden Valley (Cumbria) after William Rufus, the second Norman King of England, took the northern portion of Cumbria from the Scottish Kingdom of Alba in 1092 (Roberts 1990).

Place-names hint at ethnic complexity amongst the settlers. The *Irby* or *Ireby* names of Wirral and Cumbria show that the Irish were distinctive enough within regions already dominated by Viking settlers to attract or demand their own ethnic labels, as indeed were the Scots (Scotby, Cumbria) and possibly also the Franks (Frankby, Wirral*). Most distinctively Irish names in north-west England such as Liscard and Noctorum in Wirral (Old Irish *lios na carriage*: 'hall at the rock' and *cnocc Tírim*: 'dry hillock') occur amidst Scandinavian habitative and hybrid *-bý* names (Coates 1998).

A number of Gaelic influences on north-west England place-names have been seen as markers of possible Hiberno–Norse influence. The presence of *ærgi* names (a Norse word borrowed from the Gaelic *áirigh*) denoting summer pastures, which are most common in Scotland but by no means unknown in Ireland, has been a case in point. These names occur as *-ergh* or *-argh* in Cumbria and Lancashire, *-eary* on the Isle of Man, and *Arrowe* in Wirral.

Inversion compounds in which the personal name appears second, such as Setmurthy (Cumbria) (Old Norse *saetr*, 'shieling of Muiredach') in a word order typical of the Celtic languages, are also seen as signifiers of hybrid Norse–Gaelic influence. Personal names, where recorded historically, add to the picture. Fiona Edmonds has drawn renewed attention to two groups of lands held at the time of the *Domesday Book*, apparently by the same lord named Gilemichel (Edmonds 2009). These included estates grouped under the manor of Strickland, along the course of the Kent leading north-east from Morecambe Bay,

---

* An alternative explanation for Frankby is the Old Norse personal name *Frakki* with *-bý*.

and another tight cluster near the mouth of the Lune around Lancaster. Gilemichel (*Gille-Míchíl*) is a Gaelic name some specialists argue to be of possible Scots origin, but Edmonds argues that it is just as likely to be Hiberno–Norse name, and suggests that Gilemichel was a descendant of Dublin Vikings who obtained possession of some of the best landing-places on the Morecambe Bay coast.

It is therefore clear that place-names, parts of all of which have Scandinavian elements apparently denoting Viking settlement, are necessarily to be accepted as such without careful background and con-textual research. We are fortunate that we have the contemporary record of the *Domesday Book* of 1086 but this is not exhaustive and covers only part of the region.[*]

A considerable number of 'Viking' place-names in north-west England post-date 1100, and others are later modifications of non-Scandinavian names such as Greasby, Wirral, (appearing in the *Domesday Book* as *Gravesberie*, an exclusively Old English name). Conversely, many of the early estates estab-lished by Scandinavian lords appear to have been take-overs of existing settlements and which retained their pre-Viking names. This explains why so many parishes with important collections of 10th and 11th century Scandinavian-influenced sculpture do *not* have Norse place-names. These include English names such as Gosforth, Workington, Dearham, Brigham (Cumbria), Bolton-le-Sands, Halton and Heysham (Lancashire), Bidston, Wallasey, Woodchurch and Walton (Wirral and West Derby); or British names such as Dacre and Penrith (Cumbria). Many of these places already had pre-Viking churches with sculptural traditions. Gaelic–Norse hybrid names such as Aspatria (Cumbria) account for a much smaller number of sculpture sites. Remarkably few sites with Viking period sculpture have unambiguously Norse place-names, such as Crosscanonby and Kirkby Stephen (Cumbria) or West Kirby (Wirral).

Place-names also indicate the names of people who participated in the Viking Age society of north-west England. Although most are men—e.g., Thurstaston, Toxteth, and Ormskirk preserves Þorsteinn, Toki and Ormr—women are present too. For example in Wirral, the former Raynildes Pool and Gonnille Pool at Tranmere appear to preserve Ragnhildr and Gunnhildr (Dodgson 1972, p. 259; Harding 2002, p. 45). The importance of women and how they may have appeared in the Viking Age community in north-west England is considered by Christina Lee in Chapter 4, and the work reviewed has been inspi-rational to re-enactment groups seeking to re-create the Viking world within local communities.

Surnames recorded in later historical sources and still in use today can also assist in the search for Vikings in the British Isles in two ways. First, many people possess surnames that appear to have origins in Scandinavia such as Scholes (*skáli*—hut or hall), Holmes (*holmr*—island), and Kirk (*kirkja*—church). Kay Rogers (1991 and 1995) recorded many of these names. Modern spellings, however, do not neces-sarily reflect a name as originally bequeathed by a paternal ancestor who was a Scandinavian because patrilineal surnames as we now know them did not come into being until the 14th or 15th centuries. They merely reflect that Scandinavian or a Scandinavian-influenced dialect may have been spoken so it is almost impossible to trace a person's (paternal) line back before those centuries.

Prior to the introduction of patrilineal surnames, a person's surname usually reflected his or trade or place of birth or origin. For example we find in a 14th century document (Peet 1991) a list of inhabitants in West Lancashire contributing toward the stipend of a priest at Ormskirk that includes surnames such as de Burscogh (from Burscough), de Ellerbek (from Ellerbeck), de Kirkeby (from Kirkby), de Leyland (from Leyland), and also names like le Salter (the Salter).

Intriguingly, in Wirral we find rental records recorded in St. Werburgh's Abbey in Chester for year 1398 for Richard Hondesson, Agnes Hondesdoghter, Johanne Hondesdoghter (Great Sutton), and Mabilla Raynaldesdoghter (Childer Thornton), indicating that people still used the patronymic 'son of' and 'daughter of' conventions of surnames still used in Iceland today (Harding 2002, p. 189).

Secondly, the possession of an *old* surname (present in an area for hundreds of years but without a requirement for Viking roots) in a particular area can be used as a criterion for volunteer selection in a search for genetic ancestry in a region since surnames are also passed along the paternal line (King et al. 2006; King and Jobling 2009a, 2009b).

---

[*] *Domesday Book*'s coverage of north-west England is partial and uneven. Cheshire has a full survey that includes Lancashire south of the Ribble (Sawyer and Thacker 1987); the rest of Lancashire extending into southern Cumbria was included in the county inventory for Yorkshire (Farrer and Brownbill 1906). The northern parts of Cumbria that were not yet fully parts of England in 1086 were not covered at all.

Such a criterion was used for a recent survey of genetic ancestry in the male population of Wirral or West Lancashire where the possession of a surname that was present in those areas before 1600 was used as a requirement for participation in the survey (Bowden et al. 2008). A similar surname criterion had been applied for volunteer recruitment in a genetic survey of the Orkneys (Wilson and Goldstein 2000). The important link between surnames and genetics is considered further by Turi King later in Chapter 11.

Archaeology

Archaeology is the material record of the past in the forms of artefacts, structures, and evidence for the contemporary agriculture, landscape, and the environment. Bearing in mind that we are dealing with people whose cultural signatures changed and assimilated with those they settled amongst, we struggle at times to identify a clear Scandinavian signature in the material cultures of settlements.

Secondary indicators such as fragments of fine metalwork stolen in raids on Irish monasteries (implicitly by Vikings) become just as important as primary cultural references from the Scandinavian homeland. The great silver hoard of 7500 coins and around 1000 pieces of hack silver found at Cuerdale (Lancashire) in 1840, is emblematic of the Viking presence in north-west England (Graham-Campbell 1992, 2011), yet it contained only a minority of objects of unambiguously Scandinavian origin. Its contents show a combined monetary and bullion economy (Kershaw, Chapter 10) and their origins reflect Viking activity in Frankia, England and Ireland, along with links to markets in the Arabic world, extending as far east as modern Afghanistan.

Other contemporary silver hoards from north-west England and beyond echo this picture in less numerous but equally important ways. Directly imported Scandinavian material does exist in north-west England, for example, oval brooches from burials at Cumwhitton and Claughton Hall (Griffiths, Chapter 2; Lee, Chapter 4), but most metalwork objects including those found in burials are hybrids produced in the Viking settlements in Ireland and Britain, reflecting the convergent traditions of incoming and host cultures.

Perhaps the most distinctive object of this genre is the ringed pin, a slender bronze dress pin with a small hinged ring, of which 19 complete or partial examples were found at the Viking Age beach market of Meols in Wirral (Figure 1.14; Griffiths, Chapter 2) with further examples from Chester and from Viking burial sites at Aspatria and Cumwhitton, Cumbria.

Of course, we must remain mindful of how much evidence has not survived. Most wooden objects and architecture, for instance (that must have accounted for the vast majority of structures at the time), have decayed entirely. The glorious decorative schemes on the surviving late Viking Age wooden stave churches in Norway, at Borgund, Urnes, and Hopperstad give glimpses of what we have possibly lost in terms of Viking architecture.

Remarkably few Viking Age buildings have been documented in north-west England, and these tend to survive only as postholes and beam slots in the ground (Philpott, Chapter 7). Some structures from 10th and 11th century Chester such as the two-storey warehouse-type dwellings from Lower Bridge Street (near St. Olave's Church) may well have served the needs of Viking merchants but their detectable remains uphold few demonstrably Viking traits.

Stone sculpture presents our most graphic record of Viking culture in north-west England. Viking-period stone sculpture includes complete and fragmentary standing crosses, recumbent grave slabs, and the long-house-shaped monuments known as hogbacks that were probably grave markers although few when found were directly associated with burials. The depictions of Norse mythological scenes, most graphically at famous examples such as on Gosforth Cross 1 in Cumbria (Jesch, Chapter 3) give us vital insights into the ways in which Viking heritage was used to establish new patterns of authority and loyalty in the context of landholding and the conversion to Christianity in the 10th century.

Metal detecting, an important addition to the means of generating new archaeological evidence, has been available to the general public since the 1970s. Although illegal for use on protected monuments, metal detecting is a widely popular hobby and has almost certainly been attempted by someone at some stage on nearly all of the open farmlands, commons, and beaches across the region. The challenge that

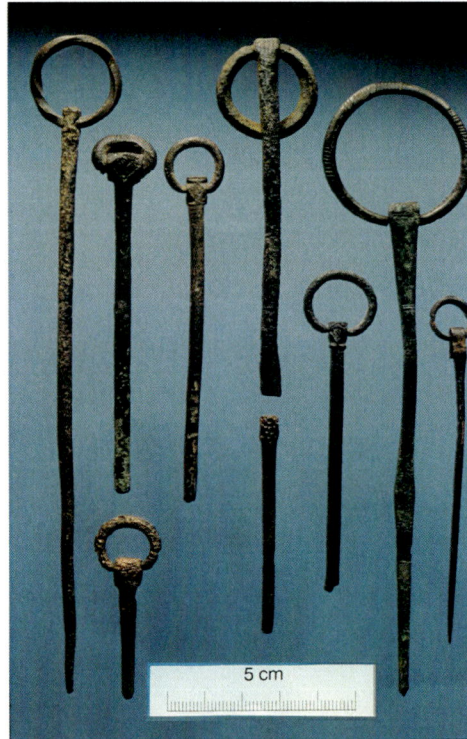

FIGURE 1.14　Ringed pins from Meols. Photo: ©Meols Project, National Museums Liverpool.

metal detecting poses to archaeological knowledge relates to the completeness and veracity of reporting on the locations and contexts of discoveries.

Until 1997, reporting facilities for metal detectorists were limited, depending on the availability of local museum staff to identify any finds presented. Few had time for anything more than cursory responses. 1997 saw the beginning of a more organised and responsive national service, the Portable Antiquities Scheme (PAS) that has reached out to metal detectorists in an attempt to create a more sympathetic and constructive dialogue.[*]

The process of establishing mutual understanding between metal detectorists and archaeologists has not always been a smooth one. The locations reported by finders have sometimes been inaccurate, sometimes deliberately so. Attempts have been made to deceive experts with fake Viking objects. A silver Thor's hammer purportedly found in Cumbria was recently dismissed as a fake.

More convincing was a worn silver penny of Anlaf Guthfrithson of York, bearing a raven (ostensibly dating from 939 to 941), allegedly found on the edge of the Dee Estuary near Neston (Wirral). The forgery was so clever that it was not possible to discern by visual means alone any reason to doubt it was genuine. It was duly noted as a Viking find on national databases and a number of scholars (the present authors included) erroneously included it as such in several publications. It took laboratory-based metallurgical analysis in 2011 by scientists at Harwell, UK using a technique known as electron probe microanalysis (EPMA) to prove that the coin could only have been made in modern times.[†]

Despite these contradictions, metal detecting has produced some spectacular and genuine results as major discoveries such as the Huxley hoard of Viking silver arm rings (2004), and the Barrow and

---

[*] http://www.finds.org.uk

[†] http://www.nottingham.ac.uk/-sczsteve/WirralNews_2Mar2011.pdf

Silverdale hoards (2011 and 2012) have been reported and their find spots investigated (Kershaw, Chapter 10). Perhaps the most archaeologically rewarding amongst these reports was the discovery by metal detecting of two composite Scandinavian oval brooches in a field near Cumwhitton, Cumbria, in 2004. This led to an excavation, uncovering a richly furnished Viking cemetery with six graves (Paterson et al. 2014; Griffiths, Chapter 2).

Metal detecting has yet to lead us to discover a fully-fledged Viking Age settlement site in north-west England. However, this is by no means an improbable prospect if metal detectorists and archaeologists work together constructively. Just such a discovery was made in Anglesey, North Wales, in the early 1990s. A report of coins and Viking hack silver found by a responsible and knowledgeable metal detectorist in a field near Llanbedrgoch and Red Wharf Bay on the eastern side of the island led the National Museum of Wales to investigate the site archaeologically.

A defended enclosure of 1.2 hectares in extent was revealed, rich in Viking Age artefacts suggesting trading activity, with buildings, burials, middens, and metal-working hearths. The research project is ongoing 20 years later (Redknap 2000; NMGW 2012). The possibilities and priorities for the archaeology of north-west England are summarised in a recent publication of an English-Heritage sponsored Research Framework (Newman and Brennand 2007): this gives guidance to curators and developers in pursuing a well-informed and research-based strategy in planning and conservation. However, chance discovery will always be a major factor, and we can never rule out a surprise.

Human Sciences and Population Genetics

Archaeology gives conditional and limited access to the human and environmental remains of the past. This is highly dependent on preservation conditions. Radiocarbon dating can be used to provide absolute dates for any organic matter encountered in the course of excavation. Although it was invented in the 1950s, its utility has only recently become widespread due to the introduction of accelerated mass spectrometry (AMS), a technique that allows a date to be obtained from a much smaller sample of material than required previously. So far, very little relevant work of this type has been carried out on material from north-west England.

Isotope analysis is now gaining prominence in archaeological research (McCarthy et al., Chapter 9). The ratio of ^{87}Sr and ^{86}Sr strontium isotopes in teeth and rib bones and a proportion of the ^{18}O oxygen isotope were used to identify Viking Age remains from an excavation at Islandbridge, Dublin (Sikora et al. 2011). Four separate warrior-style male burials in central Dublin, excavated in the early 2000s, were subjected to isotope analysis, which suggested that two were of Scandinavian origin and the other two were more likely to have originated somewhere in northern Britain (Simpson 2005; Griffiths 2010, p. 76).

A similar approach was used also to identify the remains of a woman found near Doncaster in South Yorkshire as originating from Scandinavia (Speed and Walton-Rogers 2004; Budd 2004). O-isotope analysis has been at the cornerstone of an investigation of 37 mutilated skeletons recently discovered in the grounds of St. Johns' College, Oxford (Pollard et al. 2012). All the bodies discovered were males of fighting age. O-isotope analysis has shown a strong marine food diet commensurate with raiders from Scandinavia, with one man possibly from north of the Arctic Circle. AMS dating of the bodies suggests they probably date to the 10th century and may date to the early 11th century if the effects of their fish-based diet are taken into account (known as the Marine Reservoir Effect, this factor skews dates by several years or decades too early for humans and animals with high marine protein intakes). The Oxford burials have been tentatively linked therefore to the Saint Brice's Day massacre: the killing of Danes in England on 13 November 1002 ordered by King Æthelred the Unready, which is known to have produced a massacre in Oxford.

The past two decades have seen tremendous advances in our ability to use genetic methods to probe our past. Genetic methods have now become major tools in the search for the extent of Viking ancestry of populations (see, for example, Harding, Jobling and King 2010).

Until the development of modern genetic tools, genetic analysis had been limited to the study of the distribution of 'phenotypic' characteristics, that is, the distribution of physical characteristics that are manifestations of the DNA of an individual. Such characteristics include blood group types that for years have been considered markers of population ancestry (Cavalli-Sforza et al. 1994).

Blood groups are genetically determined but unfortunately for ancestral studies are generally poorly discriminating and widespread in many populations. They are now regarded as not sensitive enough to use as markers for ancestry. Other such characteristics are skin pigmentation, stature, and facial shape, with oval faces supposedly representing people of Scandinavian origin (Geipel 1969; Figure 1.15). These features are, however, complex, poorly understood, and widely distributed across northern Europe.

Eye colour and hair colour have also been considered distinctive markers. The highest proportions of people with fair or blond hair and blue eyes appear to be found in central Sweden, Norway, Finland, the Baltic states, the northern parts of Poland, and the former German Democratic Republic with a reported 70 to 80% possessing these phenotypes (Beals and Hoijer 1965; Frost 2006; Figure 1.16). Particular physical impairments or diseases are more prevalent in Scandinavia than elsewhere in Europe but could be said to track the distribution of Scandinavian genetic influence.

One of these is Dupuytren's syndrome or digitopalmar contracture (Figure 1.17). This condition can affect people usually over 50 years of age and involves tightening of the elastic tissues in the palms of the hands, making it difficult and in some cases impossible to flex the fourth and fifth fingers. The distribution of people across Europe with this condition suggests a possible Viking origin. The Icelandic sagas even contain evidence of this connection. The *Longer Saga of Magnus of Orkney* tells about a man called Sigurdr who, after a pilgrimage to the shrine of Holy Magnus, allegedly had a complete recovery; his fingers became supple and flexible and could be put to any use (Whaley and Elliot 1993).

Although complex, the genetic bases behind all these traits are becoming much better understood (Sulem et al. 2007; Eiberg et al. 2008). For example, using only 24 targeted DNA variants, it is now possible to predict eye and hair colour; this test has already been useful in forensic casework (Walsh et al. 2013).

Although 'sporadic' cases of Dupuytren's syndrome occur, many are inherited in a simple way, suggesting a single causative gene acting in a dominant manner—if a person receives the gene for the contracture from one parent and the gene for non-contracture from the other, he or she will have the contracture. In other words, only one copy of the gene is needed to convey Dupuytren's syndrome. A recent study of one family has suggested that DNA on chromosome 16 is involved. Nonetheless, the detailed genetic cause of this condition is still largely unknown.

Because of the extra complexity of using genetic variation connected to physical differences among people, scientists commonly focus on the plentiful 'neutral' variations in DNA that have no discernible effects on a person who carries them but can provide evidence about ancestors. This technique considers characteristic features or 'markers' such as the occurrence of particular bases (G, A, T, or C) at particular locations within DNA. However, the patterns of such variants in most of individuals' DNAs are difficult to interpret because every generation undergoes reshuffling or recombination that blurs the signals of ancestry.

However, two parts of the DNA we receive from our parents are not reshuffled. Most of the DNA on the Y chromosome is passed down from father to son essentially unchanged except for rare mutations that can occur over the course of hundreds or thousands of years. Similarly, the DNA in the mitochondria is passed down the maternal line over very many generations with little or no change. Variation can occur only when there is a change or mutation and such changes are very rare. Thus by characterising the type of DNA a man has in his Y chromosome and in his mitochondria, we can obtain specific information about his paternal and maternal ancestry. With women we can also enquire about ancestry but only along the maternal line as women do not have Y chromosomes.

It is possible to characterise a person's Y chromosome DNA or mitochondrial DNA through patterns of the DNA sequence that determines his or her paternal Y chromosomal haplogroup or maternal mitochondrial haplogroup. The DNA is extracted (usually from a mouth swab taken from the inside the cheek; Figure 1.5) and is then transferred to a preservative solution. The sample is then analysed in a laboratory using a technique known as the *polymerase chain reaction* (PCR) followed by analytical procedures specific for detection of the presence of particular bases at characteristic locations.

A Dane from Jutland whose facial
features according to Geipel
"remind one irresistibly of his
forerunner, Tollund Man"

An Icelander from Reykjavik

A Norwegian woman

A German innkeeper's daughter

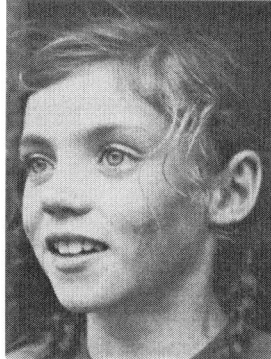

A Holstein farmer

A Swedish Lap

FIGURE 1.15 'European facial types' suggested nearly 50 years ago as a way of indicating ancestry. (*Source:* Geipel, J. (1969) *The Europeans: An Ethnohistorical Survey.* Longman, London. With permission.) Individual image captions as in original. Images reproduced with permission from Paul Popper Limited/Getty Images (top left, middle right, bottom left and right) and Mats Wibe Lund (top right). We have also reprinted (middle left) "A Norwegian woman" which was attributed to K.G. Rayson in Geipel's original publication, having been unable, despite considerable effort, to find the copyright holder. Proper acknowledgment will be made by the editors for the use of the material to the copyright-holder, if such information comes to light in any future reprint of this work.

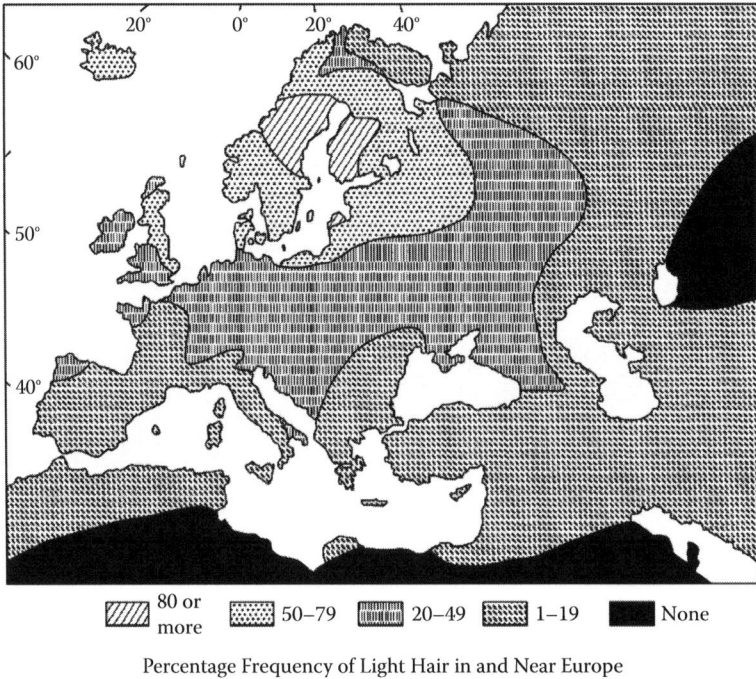

Percentage Frequency of Light Hair in and Near Europe

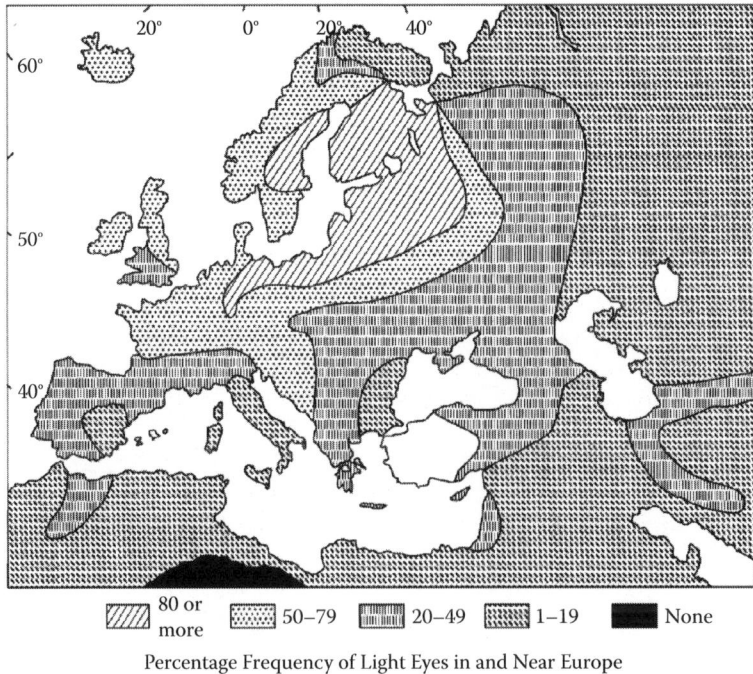

Percentage Frequency of Light Eyes in and Near Europe

FIGURE 1.16 Distribution of people with light or fair hair (top) and light or blue eyes (bottom) in Europe, showing the highest densities for both in the Baltic Sea region. (*Source:* Beals, R.L. and Hoijer, H. (1965). *An Introduction to Anthropology,* 3rd ed. Reproduced with permission from Pearson Education.)

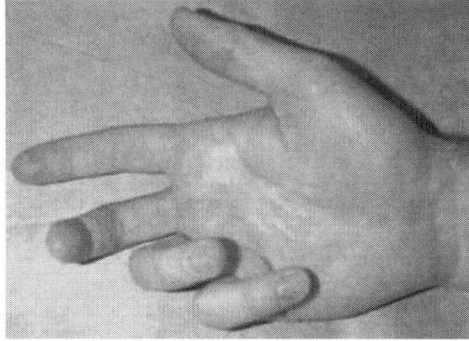

FIGURE 1.17 Dupuytren's contracture. A tightening of elastic tissue beneath the skin makes it difficult, sometimes impossible, to extend one or more of the fingers of a hand (commonly the fourth and fifth fingers). This condition is common in Scandinavia and in parts of Britain influenced by the Vikings. (*Sources:* www.med.und.edu/users/jwhiting/duprr. html and Dr. Jeff Whiting, St. Louis School of Medicine, US.)

The locations of bases allow a person's Y chromosome and/or mitochondrial haplogroup to be ascertained. These are represented by letters or letters and numbers, e.g., I1a, R1a1, KxR1* for Y-chromosomal DNA and H1, J, etc., for mitochondrial DNA. Certain haplogroups are common in modern Scandinavia and the Baltic Sea region—and regions settled by the Vikings. These include Y chromosomal R1a1 and other subhaplogroups within the general haplogroup known as K (Chapter 11, Figure 11.4), and I1a.

Sometimes the notation can be confusing to the non-specialist as the convention used has undergone several changes. For example, the notations 1, R1b, R1b3, R1b1b2, and R-M269 have all been used to represent the same Y chromosome haplogroup. Even though a common consensus appears to have been reached (Jobling et al. 2013) the notation may change again as the resolution of the method improves.

The distribution of haplogroups in different regions will differ, and one can use statistical methods to compare populations. As the resolution of the method increases, an increasing number of 'hobbyists' seek to make conclusions about their own individual DNA haplogroups based on information obtained from the many commercial testing companies now in existence although at best this can give only an idea as most people have matches that spread over a wide area† – and it only tests for one (for women) or two (for men) ancestral lines among the tens of thousands an individual might have (Thomas 2013). As a tool for comparing *populations*, the method is by contrast considerably more powerful.

The best way—if it were possible—of assessing Viking ancestry would be to analyse populations of DNA haplogroups from bones and remains of people known from isotope dating or other methods to be from the Viking Age period (Hagelberg et al. 1989) and compare them with DNA from the bones of people dated to the Viking Age in Scandinavia. However this is currently impossible because sample sizes from regions of the British Isles are small or non-existent, the DNA may be of poor quality and prone to contamination, and there may well be no descendants of such people living today.

In the future it may, nonetheless, be possible to obtain a sufficient number of samples from Viking Age human remains in Scandinavia that could provide important control data for any modern-based population study. Erika Hagelberg, Maja Krzewinska, and coworkers have been exploring the possibilities based on the Schreiner collection of the University of Oslo (Krzewinska and Hagelberg 2013).

The complementary or alternative approach is to analyse the DNA from modern people from particular areas in the British Isles and compare the distributions of haplogroups of these areas with haplogroups from different regions of Europe and beyond. This is also not without difficulties as these distributions can be unrepresentative of past populations. This concept needs methods of inference, making use, for

---

* The 'xR1' following the notation for haplogroup K means 'excluding' the large haplogroup within it known as R1; see, for example, Bowden et al. (2008) and King (Chapter 11).

† The resolution of the method is becoming increasingly more powerful as new SNPs continue to be discovered.

example, of the link between surnames and Y-DNA (King and Jobling 2009). As noted above, this link has been used to powerful effect for studies in Orkney (Wilson and Goldstein, 2000), Wirral, and West Lancashire (Bowden et al. 2008). Work is continuing across northern England following, for example, the trail of the Norse 'hogback tombstones' into North Yorkshire and the Northern Danelaw.

No modern assessment of Viking ancestry of a region would be complete without a full consideration of the genetic messages from the past. In the penultimate chapter in this volume, leading expert Turi King (Chapter 11) assesses in detail the present thinking—the balance between ancient and modern DNA analyses—describing the usefulness of how the Y-chromosome can be linked with surnames in Britain and how we can link genes with modern geography.

A detailed summary of the findings of the Bowden et al. (2008) study is also presented. Bowden found a significant pointer toward Norse ancestry with up to 50% of the total DNA mixture or 'admixture' in the sampled male populations of both Wirral and West Lancashire being Scandinavian in origin.

Since the publication of Bowden's work (2008), the resolution of the technique has improved considerably as more and many more mutations or single nucleotide polymorphisms (SNPs) are discovered; SNPs can define subhaplogroups within main haplogroups. Some particularly useful subhaplogroups and their relative distributions across Europe and Asia have been considered in two papers by Myres et al. (2011) and Busby et al. (2012). These studies appear to show, for example, a Germanic origin for the R1b subgroup known as R1b-U106 found in high levels in England but not in Wales, Scotland, or Ireland. Conversely there are high levels of the Celtic subgroups R1b-M222 and R1b-S145 in Wales, Scotland, and Ireland, but not in England. This appears to be consistent with substantial, if not overwhelming, settlement of Germanic invaders into what we now know as England during the early medieval period, and evidently addressing the debate we noted at the start of this chapter. Advances in the differential dating of the SNP mutations underlying haplogroups such as these should greatly assist historians in identifying when these migrations took place.

Indications of ancestry in terms of a person's 'whole DNA' including his or her autosomal DNA are now also possible and the *National Geographic Magazine*-based Geno 2.0 project[*] can assess the population 'group' to which a person's DNA from all his or her recent ancestors (going back several generations) is closest. Although Wirral and West Lancashire have been important test beds for a genetic search for Viking ancestry, Penrith in Cumbria (90 volunteers) was included in an earlier study (Capelli et al. 2003) along with the Isle of Man (62 volunteers), both areas with strong evidence of significant Viking presence from place-names and archaeology of significant Viking presence. Llangefni in Anglesey (80 volunteers) which had also yielded evidence of nearby Viking settlement at Llanbedrgoch (Redknap 2000).

The surname criterion was not used for volunteer recruitment in these earlier studies; only a two-generation criterion (paternal grandfather known to be from the region) was used instead. Significant ancestry was found in Penrith (approximately 37% of the DNA in the admixture of the modern population) and Isle of Man (approximately 40%) with a much smaller result (10%) in Llangefni.

Perhaps surprisingly, genetic studies in Ireland have thus far failed to find evidence of significant Scandinavian ancestry there (McEvoy et al. 2006), despite the acknowledged wealth and power of the Viking Kingdom of Dublin in the 10th and 11th centuries. Archaeological and place-name evidence for a Scandinavian presence in the Irish landscape surrounding Dublin is not unknown, but is surprisingly meagre compared to findings in the city itself (Griffiths 2010, pp. 58–59).

The Present and Future of Viking Studies

Human genetics research projects such as those described by Bowden et al. (2008) and the associated media coverage achieved prominence with the 2001 BBC2 *Blood of the Vikings* series and its popular accompanying book (Richards 2001). These events have given Viking studies a much greater public

---

[*] The Genographic 2.0 Beta Project, https://genographic.nationalgeographic.com/

profile than they attained previously. No longer merely the concerns of small numbers of academics, museum professionals, and local history societies, the Vikings have become the objects of enduring public and media fascination.

Because the most successful genetic surveys were conducted in north-west England, and notably in Wirral and south-west Lancashire (Bowden et al. 2008; Harding, Jobling and King 2010; King, Chapter 11) further raised the profiles of Viking genes and Viking heritage within the region. Major discoveries such as the Cumwhitton Viking cemetery (Cumbria) and the Huxley hoard in 2004, together with the Barrow-in-Furness and Silverdale hoards in 2011 (Kershaw, Chapter 10) have served to 'top up' and reinvigorate the general public's fascination with the ancient past, even as the previous discovery begins to fade a little from public consciousness.

Major finds from other archaeological periods, such as the highly ornate bronze Roman parade helmet found by a metal detectorist at Crosby Garrett (Cumbria) in 2010 and since sold at auction to a private buyer for over £2.3 million (despite a major public campaign to raise enough funds to buy it for the Tullie House Museum in Carlisle) further emphasise the importance and vulnerability of the region's archaeological heritage, contributing to a widespread hunger for more information.

Perhaps the most vibrant areas of Viking studies today result from the ways in which interdisciplinary research is being made available and accessible to the general public through reconstructions, exhibitions, re-enactments, and restorations of sites and monuments. A highly successful exhibition in 1990 at Liverpool Museum titled 'A Silver Saga: Viking Treasure from the North-West' gave an unprecedented platform to Viking studies in north-west England and made a focus of the Cuerdale hoard (Philpott 1990).

An associated academic conference addressed historical and archaeological issues on an interdisciplinary basis for the first time (Graham-Campbell 1992). More recently, major high profile events such as the voyage of the reconstructed *Sea Stallion of Glendalough* Viking warship from Denmark to Dublin in 2007–2008 attracted thousands of spectators, widespread media coverage, and was partnered by a major museum exhibition (Johnson 2004). Viking-themed exhibitions have become regular occurrences such as at the National Museums of Scotland (2012–2013), the National Museum of Denmark (2013), the British Museum (2014) and the Museum für Vor-und Frügeschichte, Staatliche Museen zu Berlin (2014–2015).

The restoration and re-display of stone sculptures in churches provide a number of examples of local heritage initiatives (White, Chapter 12). The restoration of the neglected collection of Viking sculptured stones at the Charles Dawson Brown Museum at St Bridget's Parish Church in West Kirby (Wirral) is a case in point. The museum created in outbuildings used by the former parish school in 1892 in memory of a local antiquarian and benefactor was closed and neglected for many years and its contents gathered dust. A restoration and re-display project generated a marked upsurge of interest in the early history of West Kirby. The re-opening took place on the weekend of 12 July 2013[*] with nearly 1000 enthusiastic locals attending (Figure 1.18).

The Chester conference that gave rise to this volume was conducted with two future aims in mind. It sought to build a consensus around interdisciplinary studies for the Viking period in north-west England and beyond. It also sought to make Viking studies as accessible and welcoming to the widest range of public and professional interests as possible. This is far from being a watering-down of academic probity. There will always be a place for education, training, and higher research in our field, combating sometimes ill-founded theories based on a lack of genuine insight. Yet the editors believe fundamentally that without well-informed popular support there can be no future for Viking studies in this region or elsewhere. We therefore present the following chapters both as contributions to interdisciplinary research and as bases for better public understanding and enlightenment.

---

[*] http://www.westkirbymuseum.co.uk/

(a)

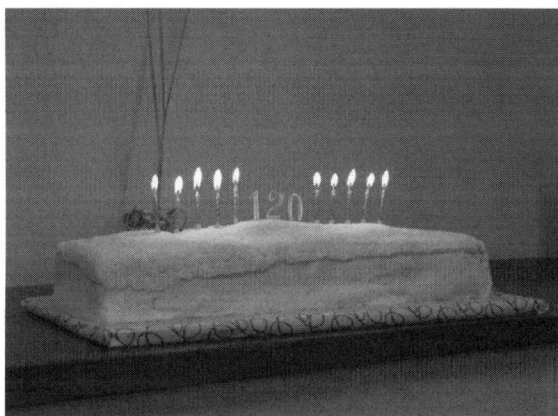

(b)

FIGURE 1.18   (a) Local people enthuse at the re-opening of the Charles Dawson Brown Museum at St. Bridget's Church, West Kirby (July 2013). St. Bridget's is the home of a significant collection of Viking Age stone sculpture including a hogback tombstone. (b) Hogback cake—a comestible replica of the St. Bridget's hogback stone made in November 2012 to mark the 120th anniversary of the opening of the museum in 1892.

REFERENCES

Abrams, L. and Parsons, D.N. (2004). Place-names and the History of Scandinavian Settlement in England, in Hines, J., Lane, A. and Redknap, M., Eds. *Land, Sea and Home: Settlement in the Viking Period*. Society for Medieval Archaeology Monograph 20, Maney, Leeds, UK, 379–431.

Armstrong, A.M., Mawer, A, Stenton, F. et al. (1950–1952). *The Place-Names of Cumberland*, English Place-Name Society XX–XXII, Cambridge, UK.

Bailey, R.N. and Cramp, R. (1988). Cumberland, Westmorland, and Lancashire North-of-the-Sands, *British Academy Corpus of Anglo-Saxon Stone Sculpture* Vol. II, Oxford University Press, Oxford, UK.

Bailey, R.N. (2009). The Sculptural Background in Cemetery, In Graham-Campbell, J.A. and Philpott, R., Eds. *The Huxley Hoard: Scandinavian Settlement in the North West*. National Museums Liverpool, UK, 24–28.

Bailey, R.N. (2010). Lancashire and Cheshire, *British Academy Corpus of Anglo-Saxon Stone Sculpture* Vol. IX. Oxford University Press, Oxford, UK.

Barnes, M. and Page, R.I. (2006). *The Scandinavian Runic Inscriptions of Britain.* Institute of Nordic Languages, Uppsala, Sweden.

Beals, R.L. and H. Hoijer (1965). *An Introduction to Anthropology* (3rd edition). Macmillan, New York.

Bersu, G. and Wilson, D.M. (1966). *Three Viking Graves in the Isle of Man.* Society for Medieval Archaeology Monograph 1, London.

Bowden, G.R., Balaresque, P., King, T.E. et al. (2008). Excavating Past Population Structures by Surname-Based Sampling: The Genetic Legacy of the Vikings in North-West England. *Molecular Biology and Evolution*, 25, 301–309.

Bradley, J., Ed. (1984). *Viking Dublin Exposed.* O'Brien Press, Dublin.

Brink, S. (2008). Who were the Vikings? In Brink, S. and Price, N., Eds. *The Viking World.* Routledge, New York, 4–7.

Budd, P. (2004). Combined O- and Sr-isotope analysis of dental tissues. *Medieval Archaeology*, 48, 61–63.

Bu'lock, J.D. (1958). Pre-Norman Crosses of West Cheshire and the Norse Settlements around the Irish Sea. *Transactions of the Lancashire and Cheshire Antiquarian Society,* 68, 1–11.

Bu'lock, J.D. (1960). The Celtic, Saxon and Scandinavian Site of Meols in Wirral. *Transactions of the Historic Society of Lancashire and Cheshire* XII, 1–28.

Bu'Lock, J.D. (1972). *Pre-Conquest Cheshire.* Cheshire Community Council, Chester.

Busby, G.B.J., F. Brisigehhi, P. Sánchez-Diaz, E. et al. (2011). The peopling of Europe and the cautionary tale of Y-chromosome lineage R-M269. *Proceedings of the Royal Society London B*, 279, 884–892.

Cavalli-Sforza, L.L., Menozzi, P., and Piazza, A. (1994). *The History and Geography of Human Genes.* Princeton University Press, Princeton, NJ.

Cavill, P., Harding, S.E., and Jesch, J. (2000). *Wirral and its Viking Heritage.* English Place-Name Society, Nottingham, UK.

Collingwood, W.G. (1927). *Northumbrian Crosses of the Pre-Norman Age.* Faber and Gwyer, London.

Collingwood, W.G. (1928). Early Monuments of West Kirby. In Brownbill, J., Ed. *West Kirby and Hilbre: A Parochial History*, Henry Young, Liverpool, 14–26 (reprinted in Cavill et al. 2000).

Coates, R. (1998). Liscard and Irish Names in Northern Wirral. *Journal of the English Place-Name Society* 30, 23–26.

Dance, R. (2013). 'Tor for to telle': Words Derived from Old Norse in *Sir Gawain and the Green Knight*. In *Multilingualism in Medieval Britain (c. 1066–1520)*, J.A. Jefferson and A. Putter, Eds. Medieval Texts and Cultures of Northern Europe. Brepols, Turnhout, 41–58.

Dodgson, J. McN. (1957). The background to Brunanburh. *Saga Book of the Viking Society* 14, 303–316 (reprinted in Cavill et al. 2000).

Dodgson, J. McN. (1972). *The Place-Names of Cheshire Part IV.* The English Place-Name Society, Cambridge University Press, Cambridge, UK.

Edmonds, F. (2009). History and Names. In Graham-Campbell, J.A. and Philpott, R., Eds. *The Huxley Hoard: Scandinavian Settlement in the North West.* National Museums, Liverpool, 3–12.

Eiberg, H., Troelsen, J., Nielsen, M. et al. (2008). Blue eye color in humans may be caused by a perfectly associated founder mutation in a regulatory element located within the HERC2 gene inhibiting OCA2 expression. *Human Genetics*, 123, 177–187.

Ekwall, E. (1922). *The Place-Names of Lancashire.* Manchester University Press, Manchester.

Farrer, W. and Brownbill, J. (1906). *Victoria History of the County of Lancaster,* Vol. 1. London.

Fellows-Jensen, G. (1985). *Scandinavian Settlement Names in the North West.* C.A. Reitzels Forlag, Copenhagen.

Frost, P. (2006). European Hair and Eye Color: A Case of Frequency-Dependent Sexual Selection? *Evolution and Human Behaviour*, 27, 85–103.

Geipel, J. (1969). *The Europeans: An Ethnohistorical Survey.* Longman, London.

Graham-Campbell, J.A., Ed. (1992). *Viking Treasure from the North West: The Cuerdale Hoard in its Context.* National Museums and Galleries on Merseyside, Liverpool, UK.

Graham-Campbell, J.A. (2011). *The Cuerdale Hoard and Related Viking-Age Silver and Gold from Britain and Ireland in the British Museum.* British Museum Press, London.

Graham-Campbell, J.A. and Philpott, R.A., Eds. (2009). *The Huxley Hoard: Scandinavian Settlement in the North West.* National Museums Liverpool.

Griffiths, D. (2010). *Vikings of the Irish Sea: Conflict and Assimilation AD 790–1050.* History Press, Stroud.

Griffiths, D., Harding, S.E., and Jobling, M.A. (2008). Looking for Vikings in North-West England. *British Archaeology* 103, 18–25.

Griffiths, D., Philpott, R.A., and Egan, G. (2007). *Meols: The Archaeology of the North Wirral Coast*. Oxford University School of Archaeology Monograph Series 68, Oxford.

Hadley, D.M. (2006). *Vikings in England, Settlement, Society and Culture*. Manchester University Press, Manchester.

Hagelberg, E., Sykes, B., and Hedges, R. (1989). Ancient Bone DNA Amplified. *Nature*, 342, 485.

Harding, S.E. (2002). *Viking Mersey*. Countyvise, Birkenhead UK.

Harding, S.E. (2007). The Wirral Carrs and Holms. *Journal of the English Place-Name Society* 39, 45–57.

Harding, S.E., Jobling, M.A., and King, T.E. (2010). *Viking DNA: The Wirral and West Lancashire Project*. Countyvise, Nottingham University Press, Birkenhead.

Harding, S.E. and Vaagan, S. (2011). *Vikingblod: spor av vikinger i Nordvest-England*. Genesis Forlag, Oslo.

Higham, M.C. (1995). Scandinavian settlement in north-west England, with a special study of *Ireby* names. In Crawford, B.E., Ed. *Scandinavian Settlement in Northern Britain*, Leicester, 195–205.

Higham, N.J. (1993). *The Origins of Cheshire*. Manchester University Press, Manchester.

Higham, N.J. (2004a). Viking-Age Settlement in the North Western Countryside: Lifting the Veil? In Hines, J., Lane, A., and Redknap, M., Eds. *Land, Sea, and Home: Settlement in the Viking Period*. Society for Medieval Archaeology Monograph 20, Maney, Leeds, 297–311.

Higham, N.J. (2004b). *A Frontier Landscape: The North West in the Middle Ages*. Windgather Press, Bollington, UK.

Jobling, M.A., Hollux, E. Tyler-Smith, C. et al. (2013). *Human Evolutionary Genetics*, 2nd ed. Taylor & Francis, New York.

Johnson, R. (2004). *Viking Age Dublin*. Irish Treasures Series, Town House, Dublin.

Kenyon, D. (1991). *The Origins of Lancashire*. Manchester University Press, Manchester, UK.

King, T.E., Ballereau, S.J., Schürer, K.E. et al. (2006). Genetic Signatures of Co-Ancestry within Surnames. *Current Biology* 16, 384–388.

King, T.E. and Jobling, M.A. (2009a). What's in a name? Y chromosomes, surnames and the genetic genealogy revolution. *Trends in Genetics* 25, 351–60.

King, T.E. and Jobling, M.A. (2009b). Founders, Drift, and Infertility: The Relationship between Y-Chromosome Diversity and Patrilineal Surnames. *Molecular Biology and Evolution*, 26, 1093–1102.

Krzewinska, M., and Hagelberg, E. (2013). The Archaeogenetics of the North Atlantic Region. In Bill, J., Ed. *Returning to Gokstad: The Art and Science of Revisiting Monumental Burials in Scandinavia*. Århus University Press.

Newman, R. and Brennand, M. (2007). The Early Medieval Period Research Agenda In Brennand, M., Ed. *Research and Archaeology in North-West England: An Archaeological Research Framework for North-West England*, Vol. 2. Archaeology North West, 73–94.

NMGW (2012). http://www.museumwales.ac.uk/en/blog/?month=2012-08

Paterson, C., Parson, A.J., Newman, R.M., Johnson, N., and Howard Davis, C. (2014). *Shadows in the Sand: Excavation of a Viking-age cemetery at Cumwhitton, Cumbria*, Lancaster Imprints 22, Lancaster.

Peet, G. (1991). Inhabitants of Ormskirk, Scarisbrick with Hurlton, Bickerstaffe, Burscough with Marton, Westhead with Lathom and Skelmersdale who promised to contribute to the stipend of the priest of the altar of Our Lady of Ormskirk, 1366. *Ormskirk District Family Historian*, 1, 2–6.

Philpott, F. (1990). *A Silver Saga*. National Museums and Galleries on Merseyside, Liverpool, UK.

Pollard, A.M., Ditchfield, P., Piva, E. et al. (2012). *Oxford Journal of Archaeology*, 31, 83–102.

Potter, T.W. and Andrews, R.D. (1994). Excavation and Survey at St Patrick's Chapel and St Peter's Church, Heysham, Lancashire 1977–1978. *Antiquaries Journal* LXXIV, 54–134.

Redknap, M. (2000). *Vikings in Wales: An Archaeological Quest*. National Museums and Galleries of Wales, Cardiff.

Richards, J.C. (2001). *Blood of the Vikings*. Hodder and Stoughton, London.

Richards, J.D. (1991). *Viking Age England*. English Heritage, Batsford.

Roberts, B.K. (1990). Late *–by* names in the Eden Valley, Cumbria. *Nomina* XIII, 25–40.

Rogers, K.H. (1991). *Vikings and Surnames*. Ebor Press, York.

Rogers, K.H. (1995). *More Vikings and Surnames*. Local History Press, Nottingham.

Sawyer P.H. and Thacker A.T. (1987). Domesday Survey. In Harris B.E. and Thacker A.T., Eds. *A History of the County of Chester*, Vol. 1. Victoria History of the Counties of England, Institute of Historical Research, London, 293–370.

Sikora, M., O'Donnabhain, B., and Daly, N. (2011). Preliminary Report on a Viking Warrior Grave at War Memorial Park, Islandbridge. In Duffy, S., Ed. *Medieval Dublin XI*. Four Courts Press, Dublin, 170–184.

Simpson, L. (2005). Viking Warrior Burials in Dublin: Is This the Longphort? In Duffy, S., Ed. *Medieval Dublin VI*. Four Courts Press, Dublin, 11–62.

Smith, A.H. (1967). *The Place-Names of Westmorland*. English Place-Name Society, Cambridge University Press, Cambridge, UK, 17–18.

Speed, G. and Walton, R.P. (2004). A Burial of a Viking Woman at Adwick-le-Street, South Yorkshire. *Medieval Archaeology*, 48, 51–90.

Sulem, P., Gudbjartsson, D.F., Stacey, S.N. et al. (2007). Genetic Determinants of Hair, Eye, and Skin Pigmentation in Europeans. *Nature Genetics* 39, 1443–1452.

Thomas, M. (2013). To Claim Someone Has 'Viking ancestors' is No Better than Astrology, *The Guardian*, 25 February. http://www.guardian.co.uk/science/blog/2013/feb/25/viking-ancestors-astrology

Townend, M. (2002). Viking Age England as a Bilingual Society, In Hadley, D.M. and Richards, J.D., Eds. *Cultures in Contact: Scandinavian Settlement in England in the Ninth and Tenth Centuries*. Brepols, Turnhout, 89–101.

Townend, M. (2009). *The Vikings and Victorian Lakeland: The Norse Medievalism of W.G. Collingwood and His Contemporaries*. Cumberland and Westmorland Antiquarian and Archaeological Society, Carlisle.

Victoria History of the Counties of England: Cumberland, 2 vols. (1901–1905); Lancashire, 8 vols. (1906–1914); Cheshire, 5 vols. (1979–2005), London.

Wainwright, F.T. (1943). Wirral Field Names. *Antiquity* 27, 57–66.

Wainwright, F.T. (1946). The Scandinavians in West Lancashire. *Transactions of the Lancashire and Cheshire Antiquarian Society* 58, 71–116.

Wainwright, F.T. (1948). Ingimund's Invasion. *English Historical Review* 63, 145–169.

Wainwright, F.T (1975). The Field-Names of Amounderness Hundred. In Finberg, H.P.R., Ed. *Scandinavian England*. Phillimore, Chichester, 229–279.

Walsh, S., Liu, F., Wollstein, A. et al. (2013). The HIrisPlex System for Simultaneous Prediction of Hair and Eye Colour from DNA. *Forensic Science International: Genetics* 7, 98–115.

Wawn, A. (2000). *The Vikings and the Victorians: Inventing the Old North in 19th Century Britain*. Brewer, Woodbridge.

Whaley, D.C. and Elliot, D. (1993). Dupuytren's disease: A legacy of the north? *Journal of Hand Surgery (British and European volume)*, 18, 363–367.

White, R.H. (1986). Viking Period Sculpture at Neston, Cheshire. *Journal of the Chester Archaeological Society*, 69, 45–58.

Wilson, E. (1979). Sir Gawain and the Green Knight and the Stanley Family of Stanley, Storeton and Hooton. *The Review of English Studies*, 30, 308–316.

Wilson, J. and Goldstein, D.B. (2000). Correlation between Y chromosomes and surnames in Orkney. *American Journal of Human Genetics*, 67, 926–935.

Winchester, A.J.L. (1987). *Landscape and Society in Medieval Cumbria*. John Donald, Edinburgh.

2

A Brief History and Archaeology of Viking Activity in North-West England

David Griffiths

CONTENTS

ABSTRACT   Viking activity is evident in historical sources, stone sculpture, artefacts, and hoards. Antiquarian accounts of burials and other discoveries have been supplemented by more recent excavations. Metal-detected material from the Portable Antiquities Scheme has become a dominant theme in recent decades. The overall amount of evidence is, however, still relatively limited and fragmentary. Source criticism is vital, and interpreting the available evidence must be conducted on a synthetic, interdisciplinary basis. Landscapes, past environments, and settlements represent continuing research problems in north-west England. Few and generally partial settlement sites have been excavated. Pointers to other chapters in this volume are included.

Historical Survey

The Viking Age in Britain and Ireland began with dramatic raids from the sea, and these remain the bases of the popular Viking image *par excellence*. The earliest recorded raids took place in the last years of the 8th century when contemporary Irish and Anglo-Saxon chronicles record attacks by heathens, northmen, or gentiles (pagans) on the monastery on Lindisfarne, Northumberland (793) and 2 years later in 795 on an Irish island called 'Rechru' (probably Rathlin, off Co. Antrim).

The earliest documented attack on English soil, more of a *fracas* than a raid, occurred in the unlikely south-western setting of Portland, Dorset, in 789 when three ships of northmen from *Horthaland* (Hordaland, western Norway) killed the King of Wessex's reeve who was presumably attempting to extract taxation from them. However, we hear very little of Vikings in north-west England until much later.

Ireland was wracked with raids from the 790s through to the 840s, when the seaborne hit-and-run method was no longer adequate to support the extent of Viking entanglement in warfare and trade and the Scandinavians founded armed winter camps on rivers known as Longphorts (*longphuirt* in Irish— ship fortresses). Longphorts at Dublin and nearby Annagassan, near Dundalk (Co. Louth), were founded in 840 and 841, with several other such sites built on rivers and coastlines shortly afterward.

The first armed bands of Danish Vikings began overwintering in England, in the Thames Estuary, from the early 850s. Further arrivals in the 860s led to the formation of a major peripatetic Danish-led heathen war band known to the English as the *Micel Here* (great army), the size of which is a matter of conjecture, but historians have debated totals ranging from several hundred to several thousand. The

Micel Here caused strife and instability across eastern England, including in 869 murdering Edmund, the King of East Anglia.

By 873, the Danes established a base on the Trent at Repton (Derbyshire), after which one part of the Viking army under Healfdene went north to capture York in 876 where, the *Anglo-Saxon Chronicle* tells us, they 'shared out the lands of the Northumbrians, and began to plough and support themselves' (Whitelock 1979, p. 195). If this statement can be interpreted as referring to the entirety of the former Northumbrian kingdom ruled from York, it would imply that Northumbrian lands west of the Pennines, including Lancashire and southern Cumbria, were similarly 'shared out' and may have had Viking land-holdings from this date onward.

The other section of the Viking army at Repton, initially under Guthrum and later Hæsten, departed south-eastward toward East Anglia, and began a long-running confrontation with the forces of King Alfred of Wessex. It seems that they made no intrusion further north or west until 893 when we are told by the *Anglo-Saxon Chronicle* that a fugitive Danish war band fleeing from defeat at the hands of Alfred's forces at Buttington (Montgomeryshire, now Powys) occupied a 'wasted [ruined] *castrum*' in Wirral—a reference to Chester (Whitelock 1979, p. 204). After a grim winter of starvation, as the *Chronicle* notes, 'deprived of the cattle and corn which had been ravaged,' they were chased out by the Mercians the following spring, initially into Wales, after which they returned by a circuitous route to East Anglia.

There were evidently few if any local supporters in the Chester region at that time were in a position to assist the tattered and battle-scarred Viking army cowering inside the ancient and ruinous walls of *Deva*. However the turning of the 9th century to the 10th was an eventful time in north-west England and the Irish Sea region, and the picture changed rapidly. In Ireland, the Viking rulers of Dublin and their followers were expelled by a coalition of Irish forces in 902, fleeing to their ships as their fortress was breached. Here we have the first genuine reference to Viking settlement in north-west England. An Irish annalistic source, known as the *Three Fragments* (or *Fragmentary Annals*) describes the exploits of a Norse-Irish chieftain (possibly a middle-ranking one) called Ingimund or Hingamund (ON *Ingimundr*) who sailed across to Wales in an attempt to secure a safer landfall, but managed only to gain a temporary foothold before continuing toward England (O'Donovan 1860, p. 235).

Remarkably, contemporary Welsh annals corroborate this by mentioning one *Ogmundr* who was ejected around 903 by the Welsh after an attempt to seize land at *Maes Osfelion* on eastern Anglesey (Wainwright 1948). According to the *Three Fragments*, upon arriving in Mercia, he sought safe landfall and sought land for the building of huts and dwellings 'near Chester' from Æthflæd (daughter of Alfred the Great) who was ruled Mercia as her husband, the Ealdorman Æthelred, was dying of disease.

The location of Ingimund's landfall in Mercia could only have been on the Wirral Peninsula, between the Dee and Mersey estuaries, as this was the only coastal territory on the Irish Sea that Mercia possessed then. After an interval, Ingimund was tempted by the riches of Chester (Figure 2.1) and the prospect of gaining better lands to betray the generosity shown by Æthflæd and persuaded his followers to mount an attack on the city. A legendary (and probably to a great extent imaginatively described) siege ensued. The Mercians tried to appeal to the honour and good nature of the Irish amongst the pagans (an interesting detail that probably reflects the prejudices of the Irish authorship of the tale), and even to split Dane from Norseman in the *melée*. The Mercians and renegade Irish threw rocks on Norsemen attempting to scale the city walls with wooden hurdles and poured boiling beer from the ramparts. The end came when the defenders let loose swarms of angry bees that stung the invaders into a humiliating retreat, although the story ends ominously with, 'it was not long until they came to wage battle again.'

We hear no more of Ingimund after the successful Mercian defence of Chester. Nevertheless the enduring Scandinavian presence on the lower Dee and Mersey, and indeed within the city of Chester, indicates that the Vikings did not go away—far from it. They established a permanent and lasting community.

The Dee–Mersey basin was subjected to a systematic militarisation by the English authorities in the early 10th century. An addendum to two versions of the *Anglo-Saxon Chronicle* called the *Mercian Register* lists a series of places accorded the status of a *burh* (initially meaning a militarised, defended enclave, but later meaning town). Chester was named as having been 'restored' as such in 907 (possibly a reference to its former Roman status).

FIGURE 2.1   Chester from the north. The 'Playing Card' plan of its Roman walls is clearly visible. (Photo courtesy of Nick Higham, University of Manchester.)

Burhs were established at Eddisbury in mid-Cheshire in 914, and at Runcorn, a prominent bluff that controlled the lowest tidal crossing point of the Mersey, in 915. On the death of Æthefæd in 918, control of Mercian affairs passed her brother, Edward the Elder of Wessex, and Mercia effectively ceased to exist as a separate political entity. Edward maintained the policy of constructing *burhs* on his north-west frontier, adding Thelwall (near Warrington) and Manchester in 919, and finally *Cledemutha* (Rhuddlan, near the mouth of the River Clwyd in north-east Wales) in 921 (Griffiths 2001).

With the exception of Chester, the north-west Mercian *burhs* were small by southern standards; some were merely temporary forts that left few permanent traces, notably Eddisbury, which was probably a very superficial and short-term re-use of the Iron Age hillfort known as the Castle Ditch, and Thelwall, which remains an unidentified site (unless it has become confused with Warrington which was later the site of a Norman castle). Manchester, along with all of Lancashire, was Northumbrian territory (Quanrud, Chapter 5), and the lands between the Mersey and Ribble in south-west Lancashire were soon added to Edward's territorial gains. Other *burhs* not mentioned in the *Mercian Register* may have extended Edward's power northward. Nick Higham argued persuasively that a *burh* was probably founded at Penwortham on the south bank of the Ribble near Preston around 920, which would have extended Edward's realm to within a stone's throw of the Viking settlements in the territory of Amounderness on the north bank of the river.

The exploits of Ingimund, despite being our only historical record of a post-902 migration from Dublin, were almost certainly not unique in north-west England and elsewhere in the Irish Sea region. Other medium-ranking leaders, some probably maintaining links with the exiled Dublin ruling clan of Ívarr, almost certainly made landfall on the coasts of Galloway, the Isle of Man, Cumbria, and Lancashire. They may have sought to join kindred settlers already in these areas prior to 902 or, like Ingimund, to establish new settlements by conquest or, equally likely, by treaty and payment. They may have come to terms with some local lords but evidently displaced others.

Two Northumbrian nobles, Abbot Tilred of Heversham (Cumbria), and Alfred, son of Brihtwulf, are recorded by the anonymous *History of St Cuthbert* as taking refuge east of the Pennines during the reign of Edward the Elder; the latter is described as 'fleeing from the pirates' (Whitelock 1979, p. 287). In a grant of 934, Amounderness ('this land of no small size,' purchased 'with no little money of my own' by Æthelstan) was given by the king to the Church of St. Peter at York, to hold 'without the yoke of hateful servitude' (a possible reference to its control by pagan hands previously). The *Chronicle of the Archbishops of York* explicitly notes that Amounderness was purchased by the king from pagans (ibid., p. 549).

This vital historical glimpse suggests that the important coastal territory in mid-Lancashire territory in mid-Lancashire comprising the Fylde Peninsula had probably been briefly under Viking rule (and we may perhaps infer similar situations in Wirral and West Derby), but even before his victory at *Brunanburh* in 937 (Cavill, Chapter 6), Æthelstan returned Amounderness to English (Northumbrian) authority—not through conquest, but by legal purchase. The interests of English and Scandinavians alike were more finely balanced and perhaps closer than we often assume.

Further north, in Cumbria, the British kingdom of Strathclyde probably have covered or at least nominally claimed the areas north of the River Eamont until 945. This weak and distant authority may have allowed some Vikings freedom to take over territory. In Cumbria, the area that became the Norman Barony of Copeland, and the Westmorland hamlet of Coupland both have Norse place-names, implying purchases (ON *kaupa land*—'bought land') that led Gillian Fellows-Jensen to suggest that 'not only great estates but small units of settlement passed into the hands of the Vikings in return for money' (1985, p. 417).

In view of the large consignments of portable wealth in the form of silver, present here in the early 10th century, most spectacularly the Cuerdale hoard found on the banks of the Ribble, it is by no means unlikely that land transfer took place through negotiation and purchase amidst a (now largely invisible) mosaic of local allegiances and disputes, buy-offs, and transient political opportunities rather than by armed conquest. There seems, perhaps surprisingly, to be virtually no demonstrable case of Scandinavian invaders wiping out and wholly redrawing existing local political boundaries after achieving total conquest and subjection of the existing inhabitants.

The three centuries of Viking activity in the Irish Sea and north-west England saw intensive cycles of warfare, settlement, trade, and assimilation with native populations—the process that became 'the story' behind the history. Armed Viking incursions into north-west England were largely held in check by the Anglo-Saxons at key battles such as Tettenhall in 910 and *Brunanburh* in 937 (Quanrud, Chapter 5; Cavill, Chapter 6), limiting the Scandinavians' scope for outright political conquest.

Independent Viking rule in north-west England was only ever a tentative prospect that briefly seemed a possibility in the first few years of the 10th century and was all but extinguished by 950, although patterns of local Scandinavian lordship remained prominent much longer in areas such as Wirral, North Lancashire, and West Cumbria.

Vikings who settled in north-west England had no alternative but to adapt to local conditions. After the mid-10th century with the onset of Christianity amongst the incomers, the settled Viking diaspora largely accommodated itself, to and built extensive familial connections amongst the existing population.

The presence of local lords with Scandinavian names in 1066 is a prominent feature of the *Domesday Book* record of Wirral and those parts of Lancashire and southern Cumbria which it covered (northern Cumbria was not covered by the book because it was not yet a part of the English kingdom; it was annexed in 1092). Most of the Scandinavian lords, such as Leofnoth and Arni, two major pre-conquest landowners in Wirral (Griffiths 2007), had been dispossessed by 1086 when the Norman takeover was recorded, and we hear no more of them.

A new power assumed control, erasing many existing allegiances, ironically one derived originally from Viking conquest in Frankia and the creation of the Duchy of Normandy. The Norman Conquest was a political watershed but did not see the end of Viking culture in the region. Later defeats of Scandinavian forces at the battles of Clontarf in Ireland (1014) and Largs in Scotland (1263) are seen by some historians as marking the end of the Viking Age in northern Britain and the Irish Sea, but in reality Scandinavian influence dissipated not in a sudden dramatic foreclosure, but in a long, slow, incremental merging into the speech, culture, and ways of life of the region.

Archaeological Overview

Place-names, often coupled with Scandinavian personal names, point toward the presence of a settled Anglo-Scandinavian rural society developing by the early to mid-10th century in north-west England.

This seems to have begun as a take-over of existing Anglian and British estates mainly near the Irish Sea coast and the trans-Pennine routeways, later expanding inland onto poorer and higher ground (Fellows-Jensen 1985).

Agriculture became more complex and geographically spread, as indicated by toponymic evidence for Hiberno-Norse shielings and upland pastures in Lancashire and Cumbria (Higham 1995). The prestige, ancestry, and changing beliefs of settled Anglo-Scandinavian landowners are marked and illustrated by their patronage of stone sculpture groups found in parish churches across the region. They were probably also amongst the most influential consumers of traded goods at the time.

Early archaeological investigations tended to raise more questions than answers. Several Viking mound burials were reported by antiquarians in the late 18th and early 19th centuries. At Aspatria, Cumbria, a burial mound was discovered and removed in 1789. About 30 m in circumference, it stood 2 m in height upon a rise known as Beacon Hill. The mound was 'levelled,' during which a stone kerb or cist was exposed beneath the cairn. It contained a poorly preserved skeleton accompanied by grave goods (now lost or locations unknown) including a sword, a 'dagger' (probably a spearhead) studded with silver, a gold or gilt strap end of Carolingian or Frankish type, a gold buckle, and iron objects including a spur, horse bit, axe, and possibly the remains of a shield.

A report of the discovery with an engraving of the finds (Figure 2.2) was published in the journal *Archaeologia* (Rooke 1792). Also illustrated were two curious carved boulders bearing circular and

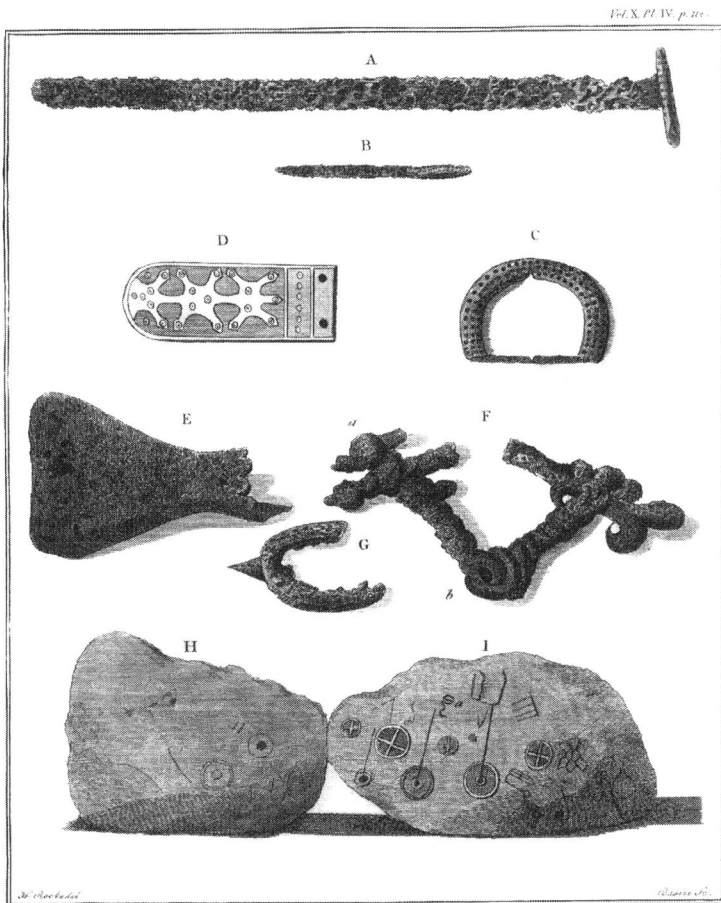

FIGURE 2.2   Aspatria burial (antiquarian drawing from Rooke 1792).

linear marks. They may have been of Bronze Age or Iron Age origin and were re-used as part of the stone cairn of the Viking burial although no other prehistoric material was discovered.

The site was re-investigated in 1997 in advance of the construction of a mobile telephone antenna on Beacon Hill (Abramson 2000). Nothing of the mound remained visible, but an elliptical pit discovered within the area investigated contained pieces of disarticulated human bone and several finds including a copper alloy ringed pin (late 9th to early 10th century), a tin-plated buckle, a folding knife, part of an axe head, and 15 very corroded iron fragments. The objects are consistent in date with those illustrated in 1792, suggesting that they were interred sometime in the early 10th century as a group.

Two further mound burials were discovered as the result of separate road realignment works in 1822. The more northerly of the two was at the inland location of Hesket in the Forest, Cumbria. A mound that caused a slight deviation of the Carlisle-Penrith road was removed and charred fragments of a burial were discovered beneath it as reported in the Newcastle journal *Archaeologia Aeliana* (Hodgson 1832; Cowen 1934 and 1948).

The materials appeared not to be inside a cist, but lay on a bed of ashes and charcoal, which led to suggestions that they had been exposed to 'considerable heat' although the heat could not have been in situ because an unburnt bone comb and case were included in the deposit together along with about 20 other items. One find was a deliberately bent sword with silver decoration on the pommel that was somewhat melted as if fired. The other metal finds were all of iron, including an axe with two spearheads (one with a deliberately bent point), a conical shield boss, a horse bit, spurs, two small iron buckles, and a small hone or whetstone.

South-east of Morecambe Bay, near the banks of the River Wyre (Lancashire), stands Claughton Hall. A new road being built to take passing traffic further away from the house necessitated the levelling of a small sandy mound, within which was a grave assemblage including two gilt copper alloy Scandinavian oval brooches arrayed back-to-back with traces of cloth, and a human molar surrounded by two beads. A sword, spear head, axe head, hammer head, and brooch converted from a gilt copper alloy Frankish or Carolingian style baldric mount were also discovered (Edwards 1970).

To the south of Claughton Hall, possible cairn burials have been noted in Lancashire at Blackrod near Wigan and at Billington on the Ribble (Edwards 1992, p. 48). References to a 'British burial mound' in the sand hills at Meols, Wirral, may be linked to the discovery in the winter of 1877–1878 of a series of iron weapons including a spearhead, a deliberately bent arrowhead, and an axe (Figure 2.5), suggesting that a furnished burial was exposed by the sea (Griffiths, Philpott, and Egan 2007, pp. 71–77). This probable Meols burial was in a highly marginal coastal location, in sand dunes only slightly above the beach—a location paralleled by Talacre and Benllech in North Wales (Griffiths 2004) and Irish grave sites such as at Three Mile Water (Arklow, Co. Wicklow) and Eyrephort (Co. Galway) (Harrison 2001).

In spring 2004, Peter Adams of the Kendal Metal-Detecting Club reported to the Portable Antiquities Scheme the discovery of first one, then two Viking-style composite copper alloy oval brooches of Viking character that date to the 9th or early 10th centuries and are the same type (Petersen Type 51) as the Claughton Hall pair. These finds (Lee, Chapter 4, Figure 4.1) occurred in open agricultural land on a small ridge above a small beck at Townfoot Farm, Cumwhitton, in the Eden Valley (Cumbria).

Finds of oval brooches, especially in pairs, normally indicate burials. This implication led to further investigation and then full excavation by English Heritage and Oxford Archaeology North during the summer of 2004 (Newman 2009; Paterson et al. 2014). The first two oval brooches came from a grave that turned out be part of a small cemetery. Six burials in all were exposed; the initial discovery was set apart just over 10 metres from the other five. The burials formed a tight cluster, orientated south-west to north-east with their heads to the west, apparently without an external boundary.

Apart from part of a badly decayed cranium found in the first grave, no human bones were retrieved due to decay caused by acidic soil conditions. Traces of wooden linings survived in two of the graves. However, a rich and varied assemblage of artefacts was recovered, as both iron and non-ferrous metals survived reasonably well. Based on the types of objects and the presence or absence of weaponry, the occupants of four of the graves were interpreted as two males and two females (including the initial discovery with the two oval brooches). The male graves included a sword, a ringed pin, a drinking horn mount, a silver-inlaid knife, a spearhead, and spurs.

There were no coins but other objects in the graves helped to give a provisional date of the early to mid-10th century for the group. Apart from the oval brooches, the objects included D-shaped and square scallop-edged belt buckles, ring-and-dot decorated strap ends, and ringed pins of types found generally in 10th century contexts elsewhere around the Irish Sea. A third oval brooch, glass beads, combs, small shears, chain links, and more utilitarian objects such as flint strike-a-lights, small whetstones, and plain iron knife blades were found in varying quantities in and around the six graves.

A number of antiquarian observations of finds (usually weapons such as swords) suggestive of further burials come from rural sites apparently not associated with settlements, mounds, or churches. These are mostly isolated finds and difficult to interpret, exemplified by a sword and a spearhead found sometime prior to 1788 during road construction near Bolton, north-west of Manchester (Graham-Campbell and Edwards 2008). B.J.N. Edwards collated a series of observations of antiquarian finds from churchyards in Cumbria including a ringed pin found beneath a church tower in Brigham, south of Aspatria, and a sword from Rampside on the Furness Peninsula (Edwards 1998, p. 21).

A burial found in the churchyard at Ormside in the upper Eden Valley in 1898 contained a sword, a shield and a small knife. This discovery occurred in the same churchyard as the earlier find (*circa* 1823) of the Ormside Bowl, a splendidly decorated Anglo-Saxon silver cup (Wilson 1984, pp. 56–59). It was originally made in the 8th century and repaired in the 9th, so it was far from new when deposited. Edwards, following the earlier comments of W.G. Collingwood, described the object as 'Viking loot' and listed it amongst stray finds of precious metals. However, its churchyard context suggests that it probably came from a burial; the rest of which was not recognised at the time of discovery.

To these older discoveries from churchyards can be added three 'transitional' groups of multiple burials with Viking period objects from ecclesiastical sites excavated under modern conditions. At St Patrick's Chapel, Heysham (Lancashire), excavations in 1977 and 1978 of up to 90 graves revealed that one skeleton had a complete, decorated, iron-riveted bone comb of 10th century Viking type placed by its pelvis (Potter and Andrews 1994). The two excavations reported in this volume add to this picture.

At Carlisle, a cemetery of 41 pre-Norman graves was discovered in 1988 during excavation prior to the construction of a new diocesan treasury at the west end of the cathedral (McCarthy et al. Chapter 9). Another group of graves with Viking period objects was discovered at the fire-gutted church of St Michael's in Workington (Cumbria) that was partially excavated prior to renovation from 1994–1997 (McCarthy and Paterson, Chapter 8).

Scandinavian lordship in north-west England is also marked by stone crosses, grave covers, and memorials. Many of the ancient parish churches in Cumbria, around Morecambe Bay, and in Wirral, house collections of stone monuments found nearby and within their bounds (the Fylde, southern Lancashire and eastern Cheshire, by contrast, have markedly fewer examples). Some areas such as Gosforth (Cumbria) have collections of such distinctive character that they have been labelled 'schools.'

Viking period stone sculptures include complete and fragmentary standing crosses, recumbent grave slabs, and the long-house-shaped monuments known as hogbacks that were probably grave markers although few when found were directly associated with burials (Lang 1984). Art and inscriptions are particularly valuable for charting changing cultures and beliefs, and in the case of the stone monuments, their territorially rooted and publicly visible character maintains vital links to place, landscape, and artistic patronage on the part of local rulers.

A famous example of a graphic combination of Norse legend and Christian symbolism is formed on the shaft of the slender and miraculously well-preserved sandstone cross still standing in the churchyard at Gosforth (Cumbria) (Jesch, Chapter 3, Figure 3.4). We see a group of images representing the Viking Doomsday, Ragnarók, interspersed with scenes of the Crucifixion. The Norse gods Viðar and Loki are depicted, the former on the east face fighting a terrible monster that probably represents Fenrir, the wolf that killed Oðin, while on the west face, the latter is being punished for his role in the death of the virtuous god Baldr.

Other mythological scenes are visible on crosses from Halton (Lancashire) (Jesch, Chapter 3, Figure 3.3) and at sites in the Viking-settled areas of Yorkshire and the Isle of Man. The use of Norse and pagan imagery may seem strange on what are explicitly Christian crosses. However, it seems these were used as a means of drawing together older traditions and forming cross-cultural allegiances to underpin the landed power of patrons in the new Christian Age (White, Chapter 12).

FIGURE 2.3   Bidston Hogback, Wirral. (Photo courtesy of Ross Trench-Jellicoe.)

Hogbacks seem, on the basis of their frequency east of the Pennines, to be 10th century innovations and closely linked to the kingdom of Northumbria and its lands west of the Pennines in Cumbria and North Lancashire, and also in Wirral. Various sub-types exist in size and decoration; the Cumbrian examples are generally thinner and taller than those found elsewhere.

Most distinctive are those of the Brompton School (named after a parish church in North Yorkshire with a particularly splendid group), with gripping beasts or bears at the ends of the 'long houses.' A miniature version of this Northumbrian type was recently unearthed from the garden of a former parish vicarage at Bidston, Wirral (Bailey and Whalley 2006) a parish perhaps significantly dedicated to a Northumbrian saint, St. Oswald (Figure 2.3).

Unlike the crosses, hogbacks rarely exhibit well-understood mythical scenes and present a more eclectic range of images ranging from battle scenes to beasts. A battle scene with helmeted standing warriors arrayed in formation with round shields and spears appears on a Gosforth hogback ('The Warrior's Tomb') and a ship in full battle formation arrayed with shields appears on a hogback from Lowther (Cumbria). Perhaps this difference indicates that, like the furnished graves, hogbacks were primarily personalized statements, and perhaps not intended for the same publicly didactic purposes as crosses.

Hoards and Trade

The discovery of a huge Viking silver hoard at Cuerdale on the banks of the Ribble in 1840 (Chapters 1 and 10, this volume) produced a flurry of interest from antiquarians and collectors but little consolidated research at the time. Although the hoard has now been described in full (Graham-Campbell 2011), we still have a very limited understanding of the precise context of its discovery in archaeological terms.

Due to the manner in which the hoard was dispersed after discovery, it is impossible to put an exact figure on its contents, but James Graham-Campbell stated that it contained 'some 7500 coins, together with over 1000 ingots, rings and hack silver pieces.' The silver was already spread into separate barrow loads of soil and various people's pockets almost as soon as it was noticed, but fragmentary remains of a lead container and five bone pins were also retrieved, suggesting the items had been buried together in moneybags within a metal box or chest.

The hoard can be dated between 905 and 910 based on the latest coin issues—a time of particular resonance in the turbulent history of the region (Quanrud, Chapter 5). Around 5000 of the coins were Viking issues of York and East Anglia, together with about 1000 Anglo-Saxon coins of Alfred and Edward the Elder, and a similar number of Frankish coins from the Rhineland and Loire areas along with Italian and Byzantine issues. A minor portion of the coins come from further afield, demonstrating the extent of Viking trading links. Four have been attributed to the trading town of Hedeby in southern Denmark, and around fifty Arabic (Kufic) coins range in mintage from Al Andalus in Islamic Spain to Al Banjhir in the Himalayan foothills.

Cuerdale is the most spectacular of a series of silver hoards deposited on the Irish Sea coastline of north-west England in the early decades of the 10th century. The earliest recorded of these was discovered in April 1611 at the Harkirk, a recusant burial ground near Little Crosby (formerly south-west Lancashire, now Merseyside). A hoard of about 300 coins, the deposition of which probably occurred around 910, was discovered in the enclosing ditch and described and illustrated by landowner William Blundell. His notes and a copper engraving of 35 of the coins survive, although the silver has been lost. Kufic coins are represented in another early coin hoard find in the mid-18th century at Dean, near Cockermouth (Cumbria). A mixed hoard containing ingots, hack silver and around 100 Anglo-Saxon coins deposited around 935 or 940 was found at Scotby, east of Carlisle, in 1855.

The spread of metal-detecting activity in recent years added several new examples to those found in earlier decades and centuries. A hoard of 21 Hiberno-Viking arm rings and an ingot found at Huxley near Chester in 2004 was deposited within the same few years at the start of the 10th century as the Cuerdale hoard and was also found with the remains of a lead container.

A hoard of hack silver, bent bars, a section of an ingot, and three Kufic coins was found in 1997 at Tewitfield, Warton, near Carnforth, on the north-eastern margins of Morecambe Bay. In 2002, a small hoard containing an ingot, a bar, a polyhedral blob of silver, and a pierced stone were discovered beside the River Dee south of Chester at Eccleston.

Chester has produced four hoards, the earliest of which is a coin-only hoard found at St John's Church in 1862 and dated to 917 or 920. A large mixed hoard dating to 965 or 970, with 27 ingots, 120 pieces of hack silver, and 547 coins in a broken Chester-ware pot was found at the Castle Esplanade, Chester, in November 1950 (Thacker 1987). This was an unusual hoard with a very long age-structure (it contained many obsolete coins going back to Alfred's reign along with the most recent issues used to determine the date of deposition). The hoard is less like mid- and later 10th century hoards in England and more like contemporary hoards in Ireland and the Isle of Man.

A stone ingot mould was found near the Castle Esplanade at Cuppin Street in 1986, which suggests that the hoard may have been associated with silver working area or even the Chester mint, and was possibly a silver worker's or moneyer's private reserve of scrap silver that would have been intended for the melting pot. Two more Chester hoards from Eastgate Row and Pemberton's Parlour (found in 1857 and 1914, respectively) date from the 970s, and their coin-only contents reflect a swing away from the use of bullion. Hoards have continued to be found in the region with surprising regularity. Two major discoveries by metal detectorists in 2011 at Barrow-in-Furness and Silverdale, both on the margins of Morecambe Bay, are discussed by Jane Kershaw in Chapter 10.

Hoards cannot tell us everything we wish to know about patterns of wealth and exchange. Single finds of silver objects and coinage that may have been dropped by mistake during a transaction perhaps convey more readily the actual patterns of use of silver and coins. Occasionally, individual items of Viking silver are found in coastal locations, such as a tapering rod arm ring with twisted ends found on the edge of the Solway Firth on the English–Scottish border near Gretna in the 1970s.

Metal detecting activity has contributed considerable new information to our understanding of the spread of Viking period finds. Recent single finds of Scandinavian objects such as a gold ring found 'near Kendal,' a decorated lead weight found near Preston (Lancashire) (Kershaw, Chapter 10, Figure 10.8) and an open-work copper alloy scabbard chape from Chatburn (Lancashire) (Edwards 1998, p. 29) enhance the range evidence hitherto dominated by antiquarian finds.

Finds of objects from coastal locations suggesting early Irish Sea Viking activity include two items of Irish-style gilded copper alloy metalwork, one of which is a human-faced escutcheon from Arnside on the shores of Morecambe Bay (Youngs and Herepath 2001; Youngs 2002). These add to earlier finds of Irish-style metalwork including an interlace-decorated boss from the Ribchester Roman fort (Thompson-Watkin 1883) destroyed in the 1941 bombing of Liverpool Museum (see Meols, below).

Trading activity is easier to detect when numerous discoveries occur at a single location, such as at Meols (Wirral). Meols (name from ON *Melr* or sand hills) first became known in the 19th century when the sea began to erode a sand spit known as Dove Point. The spit formerly protruded about 500 m into

FIGURE 2.4   Meols from the east, showing the location of the now-vanished Dove Point. (Photo courtesy of Robert Philpott, National Museums Liverpool.)

the Irish Sea from the present northern coast of the Wirral Peninsula (Figure 2.4). As the sea removed the overlying sand dunes, complex archaeological layers dating between the Mesolithic and post-Medieval periods were exposed. We would know little today about Meols but for the devoted attention of a group of antiquarians from 1817–1905.

Foremost amongst them was the Reverend Abraham Hume (1814–1884). His monograph *Ancient Meols*, of 1863, is a classic of Victorian scholarship but came too early to tell the whole story of the Meols landscape which continued and continues to reveal archaeological evidence. Thanks to Hume and others, much of the material discovered at Meols was collected and saved, ultimately to enter museums in Liverpool, Chester, Warrington, and London. Bomb damage to Liverpool Museum in May 1941 caused much disruption, and it took until 2007 for a comprehensive study of the site and collections to be published (Griffiths, Philpott, and Egan 2007).

Finds at Meols of imported metalwork, ceramics, and coinage of the Iron Age, Roman era, and pre-Viking periods indicate that sporadic trade was conducted on the sandy foreshore there for many centuries before the Vikings appeared. Coincident with the time of the Norse settlement in Wirral following the arrival of Ingimund and the rise of nearby Chester as a trading port, Meols experienced an upsurge in trading activity. Amongst the other finds from Meols are several objects suggestive of Viking raiding and trading activity in the 9th or 10th centuries (Figure 2.5).

A small gilded copper-alloy plaque re-used as a strap end is probably an Irish-made object of the 8th century (perhaps a book mount). Nineteen complete and incomplete parts of copper alloy ringed pins of later 9th to early 12th century types are identifiable. The number of ringed pins exceeds the finds at York and is outnumbered only by the number of finds from Dublin (the only place where we know they were manufactured). A drinking horn terminal (now lost), a small six-sided pyramidal bell, a stirrup mount, and a small buckle decorated in derivative versions of early 11th century Ringerike style constitute further evidence of contacts with Dublin and the Danelaw (Griffiths, Philpott and Egan 2007, pp. 58–77).

A small copper alloy bird with suspension loops above and below resembles a similar decorative bird from a balance scale found on the island of Gigha (Argyll) (Grieg 1940, pp. 29–30); see Figure 2.6. Similar examples are also known in Scandinavia. The discovery of the bird mount at Meols adds to the evidence for trade at the site.

Meols probably acted as a seasonal beach market where Cheshire salt, slaves, livestock, lead from mines in Flintshire, possibly copper from the North Wales coast, and portable domestic and dress items were traded and exchanged. The position of Meols on the outermost margins of Mercia nearest to deep water is central to the network at the eastern end of what became a regular Viking trade route between

FIGURE 2.5 Viking weapons (axe, shield boss, and deliberately-bent spearhead) found in 1877 and 1878 from a possible pagan grave at Meols, Wirral. See Griffiths, Philpott, and Egan, 2007, p. 76. (Copyright National Museums Liverpool.)

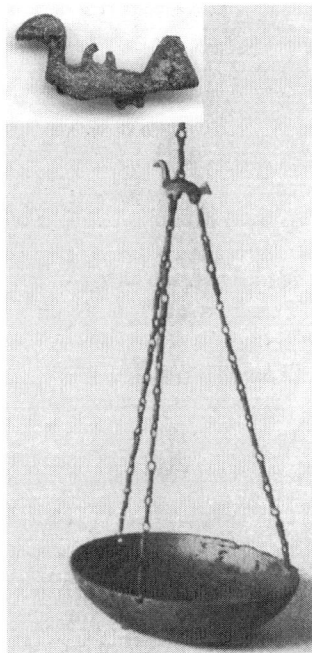

FIGURE 2.6 Bird mount from a balance scale from Meols with comparable complete example from Jåtten, Rogaland, Norway. (Petersen 1940)

Dublin, Anglesey, and Liverpool Bay. Meols may have had an illicit role in evading taxes and dues based on its location just outside the fiscal remit of the port of Chester as defined in medieval documents that make it clear that 'Arnald's Eye,' a sandstone reef at the mouth of the Dee, was the outer limit of the port.

Perhaps Meols was protected as a separate trading zone by the semi-independence of the Viking settlements in Wirral. There was also a permanent settlement there. Antiquarian records of the 1870s through the 1890s clearly indicate that buildings were being eroded from under the sand dunes. Extensive evidence of later medieval structures and circular round houses can probably be rejected as relevant to the Viking period, but descriptions of 'lines of wattle' forming cattle sheds and nearby stone- and clay-walled buildings (approximately 10 × 4 m in size) are strikingly reminiscent of Viking long house-style buildings excavated elsewhere such as at Llanbedrgoch, Anglesey (Redknap 2004).

The local influence of Meols is marked by finds of coinage and other traded items in its hinterland and upriver toward Chester, such as at Moreton and Irby (Philpott, Chapter 7). A small silver ingot of Viking style was found in a field at Ness. Nearby, in the neighbouring village of Puddington, a Frankish denier of Charles the Bald (840–877) minted at Melle was discovered in 1993 (Cowell and Philpott 1993). These Wirral find spots have yet to be investigated further and their locations may well hold further evidence for rural settlement.

As studies of the Danelaw have found (Stocker 2000), the distribution of stone sculpture sites is highly significant in relation to trade and markets. A good example of this is the pattern of parish churches with stone sculptures in North Wirral, at West Kirby, Greasby, Woodchurch, Bidston, and Wallasey. Together they form a semi-circular cluster at the centre of which lies Meols. The place-name Thingwall, found both in Wirral and West Derby, denotes the field of a local assembly site or mound, in a modest fashion not unlike the Norse *Þing-vollr/vellir* assemblies of Iceland and the Isle of Man.

The Wirral Thingwall site has often been identified with Cross Hill, a low rise near the edge of Thingwall township in central Wirral. Cross Hill has a good view of the peninsula, extending to the Mersey estuary to the east. A low mound may have existed on top, but has been ploughed down to a mere trace. Recent landscape research by Dean Paton revealed two convincing alternative sites, the former site of Thingwall Mill (on a prominent mound now levelled) and a nearby field called Dale Heaps—another reference to mounds. Dale Heaps is significantly at the intersection of three local boundaries (Paton 2011).

A parallel assembly site on the Lancashire side of the Mersey at Thingwall Hall near Roby has been built over. The hall stands on a terraced rise, which is, however, more likely to be a Victorian garden earthwork than a Viking mound.

With good historical, place-name and artefactual evidence for a Viking presence in north-west England, it might be assumed that there would also recognisable archaeological evidence of settlement in the form of buildings, enclosures, and field systems. This has, however, proven a stubborn problem for archaeologists to solve. Excavations of settlement sites have continued sporadically throughout the late 20th and early 21st centuries. Individual site works at Irby and Moreton, Wirral (Philpott, Chapter 7), and Nick Higham's work at Tatton, Cheshire have begun to fill out what had hitherto been a very sparse picture for rural settlement.

At a number of excavations at rural upland settlement sites in Cumbria in the Duddon Valley (Figure 2.7), finds and fragmentary remains of medieval long houses may hold clues of Viking presence.

FIGURE 2.7 Excavation of medieval long house arguably derived from Scandinavian building tradition at Stephenson's Ground, Scale, in the Duddon Valley, Cumbria. (Copyright Lake District National Park.)

However, to date, several of these sites have produced material dated much later than the Viking period (http://www.duddonhistory.org.uk/downloads/longhouses.pdf).

Tantalising radiocarbon dates of the 10th and 11th centuries have been published in interim reports from settlement sites such as Bryant's Gill (Dickinson 1985) and a corn-drying kiln from Ewanrigg, Cumbria (Bewley 1987), but a consolidated picture based on final and full publication remains elusive. Aerial photographic campaigns such as those undertaken over many years by Nick Higham in Cumbria and Robert Philpott in Cheshire, Merseyside, and South Lancashire, have revealed some hints of possible sites, generally to be seen against a background of multi-period activity.

Large regional palaeoenvironmental syntheses such as the North-West Wetlands Survey (Cowell and Innes 1994) have cast only oblique and partial light on this period. Geophysical surveys have made limited impacts to date despite their long-standing importance in landscape studies elsewhere. The impacts of newer remote-sensing techniques such as LiDAR on the Viking period archaeology of north-west England are likewise at an early stage, but promise much for the future.

Towns and Urban Economies

The former Roman fortress at Carlisle was probably at least partly reoccupied during the Viking period, but apart from the cemetery discussed here (McCarthy et al., Chapter 9) we know very little about its extent and topography as a settlement. The only significant urban centre in the north-west from which we have significant archaeological evidence of the Viking period is Chester. From around 907 it possessed a mint, defences, a harbour, and a series of important churches. Its status was confirmed in 973 when the *Anglo-Saxon Chronicle* records that English King Edgar received the submission of six (or possibly eight) *subreguli* (sub-kings) from the Celtic kingdoms of northern and western Britain, at least three of whom may have been of Viking origin (Whitelock 1979, p. 228).

St John's Church, to which we are told by John of Worcester that Edgar was rowed in a boat by the *subreguli*, stands outside the walls next to the site of a Roman amphitheatre and has a significant collection of 10th and 11th century red sandstone sculptures including six circle-headed crosses.

Chester's brief occupation by Danish Vikings in 893 and 894 and its reoccupation by the Mercian English (becoming a *burh* in 907) took place amidst and on top of large amounts of derelict Roman masonry rubble from ruined barracks, official buildings, and storehouses, but with the fortress walls and street plan still semi-intact and discernible.

The 10th century Mercian defences of Chester were therefore partly-adapted from those of *Deva*, although the western and southern walls of the Roman 'playing card' plan were probably superseded by extensions from the northern and eastern parts of the circuit to the River Dee. The excavation at 26-42 Lower Bridge Street in 1974–1976 outside the southern Roman wall but within the larger post-907 enclosure near the harbour and bridge has produced the most complete and informative series of building remains from the 10th and 11th centuries in Chester (Mason 1985).

Phase IV of the site, dated to the early to mid-10th century, produced evidence of three sub-rectangular timber buildings built over rock-cut cellars, with sloping entrance ways measuring 5.10 × 4.5 m and 1.8 m in depth. Their occupation may have lasted somewhat longer than the excavator suggested as they may have been modified for further occupation toward the end of the 10th century and into the 11th, after which they were re-used as tanning pits.

These buildings are similar to contemporary Anglo-Saxon urban buildings in London, Canterbury, and Oxford and later 10th century plank-constructed buildings in Coppergate, York, and Thetford. They resemble four sunken-featured buildings of later 11th century date excavated in the Hiberno-Norse town of Waterford, in 1987 and 1988 (Hurley and Scully 1997).

In Chester, these represent a type apparently restricted to the newly built-up area between the Roman walls and the river. Contemporaneous structures within the Roman walls were revealed in excavations in

3 cm

FIGURE 2.8   Viking disc brooch from Chester (right) with a similar example from Dublin. (Copyrights Cheshire West, Chester Council, and National Museum of Ireland.)

the 1970s and 1980s, at Crook Street and Hunter's Walk in the north-west quadrant of the Roman fortress reveal partial survivals of simpler, ground-based post-built hall-type structures. In the north-eastern quadrant, an open-sided shed, gravel path, corn-drying kiln, and antler-soaking pit were found at Abbey Green, in 1975–1978 (Ward 1994).

The most impressive individual artefact of Viking type from Chester is a copper alloy disc brooch found in excavations at Hunter Street School in 1981 (Figure 2.8). Measuring 3.2 cm in diameter, it consists of two separately cast discs (with the rear plain disc bearing the pin). On its convex openwork face plate is a double-contoured ribbon animal decorated with transverse billets and spiral hips coiled upon itself within the circular border. The design encompasses elements of Viking art styles of the late 9th to mid-10th centuries, and thus is probably an early to mid-10th century piece. It is paralleled by a brooch from High Street, Dublin and other examples from Cottam near York, East Anglia, and Iceland. A complete copper alloy ringed pin of the polyhedral-headed type (dated by the Dublin corpus to the 10th century) was found in excavations at Crook Street in 1973 (Ward 1994). At least three other ringed pins or parts thereof were found in the city. A ring, possibly from a ringed pin, found at Northgate Brewery is made of silver, and a stray find of a plaited gold finger ring also points to Viking connections.

Moneyers' names on early 10th century coins of the Chester Mint include the Scandinavians Oslac and Thurstan. The presence of Scandinavians can be glimpsed in the material culture of a mixed trading population. Minor names in the city such as the now-superseded Clippe Gate and Wolfield Gate (from the Old Norse personal names *Klyppr* and *Ulfhildr*) also betray their presence.

Alan Thacker drew attention to the fact that Handbridge, a suburb across the river from the southern edge of the city, was assessed in the Domesday Book in the Anglo-Scandinavian style of carucates rather than in English hides. A reference in Domesday to the city having twelve justices is a feature more commonly found in the Danelaw (Thacker 1987 and 1988).

Some commentators have seen the city's famous timber-framed two-storey shopping galleries known as the Rows as Scandinavian-influenced features, quoting the (not very close) analogy of the Bryggen warehouses in Bergen (Norway), although the wooden fabric of the Chester Rows has not been dated any earlier than the 13th century, well after the Viking period.

Church dedications, including the dedication to St Olave (Olaf) on Lower Bridge Street and the nearby Hiberno-Norse dedication to St Bridget confirm the place of Scandinavian and Irish Sea Viking culture in the city's historic legacy. These topographical features in the cityscape remained *in situ* throughout medieval times and later as symbolic references to Hiberno-Norse ancestry, mercantile trading links, and historical prestige of the citizenry.

Similar importance can be accorded to rural parishes, church dedications, and local traditions. As research develops and changes, the preoccupations of former generations of archaeologists with structure, topography and dating are being expanded by studies of human inhabitation, experience, emotion, and embodiment (Hadley and ten Harkel 2013). Beyond the mere anonymous demography of populations, the lives and experiences men, women, children, the elderly, sick, unfree, and socially excluded, are now being studied and written into archaeological interpretation.

Conclusion

The principal feeling arising from wrestling with the evidence for Vikings in north-west England could be categorized as frustration coupled with continued fascination. We remain unsure about so many aspects of the period. When and where did the first visitation of Vikings occur in the region? Was the visit war-like or peaceful? When and where were the first permanent settlements? How many settlers came, and how did they interact with the existing populations? Were there any common patterns or did situations develop differently in separate areas, perhaps even from one lakeland valley or lowland parish to another?

The stock of data is highly partial but remains tantalizing. Compared to W.G. Collingwood and his generation of scholars, we benefit not from any greater powers of insight or understanding. Their observations were at least as effective as any since—but we benefit from immeasurably faster and more effective technology in conducting our research. We also have the advantages of an enormously expanded literature, and most obviously, another century and more of discovery and excavation.

No one can predict the likelihood of future archaeological discoveries, but experience shows that they will continue to appear. If the momentum of the early years of the 21st century is maintained, new discoveries are likely to be extremely impressive and newsworthy. Similarly, as scientific techniques advance, we may expect the quantity and quality of data from isotopes, genetic mapping, and scientific dating to improve beyond the imaginations of the early pioneers. The picture in 30 or 50 years may have changed beyond recognition.

REFERENCES

Abramson, P. (2000). A re-examination of a Viking Age burial at Beacon Hill, Aspatria. *Transactions of the Cumberland and Westmorland Antiquarian and Archaeological Society,* C, 79–88.

Bailey, R.N. and Whalley, J. (2006). A miniature Viking-Age hogback from the Wirral. *Antiquaries Journal* 86, 345–356.

Bewley, R.H. (1987). Ewanrigg. *Current Archaeology* 103, 232–233.

Cowell, R. and Innes, J. (1994). The wetlands of Merseyside. *North West Wetlands Survey* 1, Lancaster University Press, Lancaster, UK.

Cowell, R. and Philpott, R. (1993). Some finds from Cheshire reported to Liverpool Museum. *Cheshire Past* 3, 10–11.

Cowen, J.D. (1934). A catalogue of objects of the Viking period in the Tullie House Museum, Carlisle. *Transactions of the Cumberland and Westmorland Antiquarian and Archaeological Society,* Series 2, 34, 166–187.

Cowen, J.D. (1948). Viking burials from Cumbria. *Transactions of the Cumberland and Westmorland Antiquarian and Archaeological Society*, Series 2, 48, 73–76.

Dickinson, S. (1985). Bryant's Gill, Kentmere: Another Viking period Ribblehead? In J.R. Baldwin and I.D. Whyte, Eds., *The Scandinavians in Cumbria*. Scottish Society for Northern Studies: Edinburgh, pp. 83–88.

Edwards, B.J.N. (1970). The Claughton burial. *Transactions of the Historic Society of Lancashire and Cheshire* 121, 109–116.

Edwards, B.J.N. (1998). *Vikings in North-West England: The Artifacts*. Centre for North-West Regional Studies, Lancaster, UK.

Fellows-Jensen, G. (1985). *Scandinavian Settlement Names in the North-West*. C.A. Reitzels Forlag, Copenhagen.

Graham-Campbell, J.A. (2011). *The Cuerdale Hoard and Related Viking-Age Silver and Gold from Britain and Ireland in the British Museum*. British Museum Press, London.

Graham-Campbell, J.A. and Edwards, B.J.N. (2008). An 18th century record of a Lancashire Viking burial. *Transactions of the Lancashire and Cheshire Antiquarian Society* 104, 151–158.

Griffiths, D. (2001). The North-West Frontier. In D. Hill and N. Higham, Eds., *Edward the Elder, 899–924*. Manchester University Press, Manchester, pp. 167–187.

Griffiths, D. (2004). Settlement and acculturation in the Irish Sea region. In J. Hines, A. Lane, and M. Redknap, Eds., *Land, Sea and Home: Settlement in the Viking Period*. Society for Medieval Archaeology Monograph 20, Maney, Leeds UK, pp. 125–138.

Griffiths, D. (2007). Maen Achwyfan and the context of Viking settlement in North-East Wales. *Archaeologia Cambrensis* 155, 143–162.

Griffiths, D., R.A. Philpott, and G. Egan (2007). *Meols: The Archaeology of the North Wirral Coast*. Oxford University School of Archaeology Monograph Series 68, Oxford, UK.

Griffiths, D. (2010). *Vikings of the Irish Sea: Conflict and Assimilation, AD 790–1050*. History Press, Stroud, UK.

Hadley, D.M. and ten Harkel, A., Eds. (2013). *Everyday Life in Viking-Age Towns: Social Approaches to Towns I England and Ireland circa 800–1100*. Oxford: Oxbow.

Harrison, S.H. (2001). Viking graves and grave goods in Ireland. In A.C. Larsen, Ed., *The Vikings in Ireland*. Roskilde, pp. 61–75.

Higham, M.C. (1995). Scandinavian settlement in north-west England, with a special study of *Ireby* names. In B.E Crawford, Ed., *Scandinavian Settlement in Northern Britain*. Leicester: Leicester University Press, pp. 195–205.

Higham, N.J. (1992). Northumbria, Mercia, and the Irish Sea Norse. In J. Graham-Campbell, Ed., *Viking Treasure from the North-West: the Cuerdale Hoard in its Context*. National Museums and Galleries on Merseyside: Liverpool, pp. 21–30.

Higham, N.J. (2004a). Viking-age settlement in the north-western countryside: lifting the veil? In J. Hines, A. Lane, and M. Redknap, Eds., *Land, Sea and Home: Settlement in the Viking Period*. Society for Medieval Archaeology Monograph 20, Maney, Leeds, pp. 297–311.

Higham, N.J. (2004b). *A Frontier Landscape: The North West in the Middle Ages*. Windgather Press, Bollington, UK.

Hodgson, C. (1832). An account of some antiquities found in a cairn near Hesket-in-the-Forest, Cumberland. *Archaeologia Aeliana* 2, 106–109.

Hume, A. (1863). *Ancient Meols or Some Account of the Antiquities Found near Dove Point on the Sea Coast of Cheshire*. John Russell Smith, London.

Hurley, M.F. and Scully, O.M.B. (1997). *Late Viking Age and Medieval Waterford Excavations, 1986–1992*. Waterford Corporation, Ireland.

Lang, J.T. (1984). The hogback: a Viking colonial monument. *Anglo-Saxon Studies in Archaeology and History* 3, 86–176.

Mason, D.J.P. (1985). *Excavations at Chester, 26-42 Lower Bridge Street: The Dark Age and Saxon Periods*. Grosvenor Museum Archaeological Excavation and Survey Reports 3, Chester.

Newman, R. (2009). The Cumwhitton Viking cemetery. In Graham-Campbell, J.A. and Philpott, R., Eds., *The Huxley Hoard: Scandinavian Settlement in the North West*. National Museums Liverpool, pp. 21–23.

O'Donovan, J., Ed. (1860). *Annals of Ireland: The Three Fragments*. Dublin.

Paterson, C., Parsons, A., Newman, R. et al. (2014). *Shadows in the Sand: The Excavation of a Viking-Age Cemetery at Cumwhitton, Cumbria*. Lancaster Imprints 22, Lancaster, UK.

Paton, D. (2011). *Things aren't always what they seem: a critique of recent approaches to early medieval assembly sites in Britain with a case study focusing on Thingwall, Wirral*. Unpublished BA dissertation, University of Chester.

Petersen, J. (1940). British antiquities of the Viking period found in Norway. In H. Shetelig, Ed., *Viking Antiquities in Great Britain and Ireland*. Part V, Aschehoug, Oslo.

Potter, T.W. and Andrews, R.D. (1994). Excavation and survey at St Patrick's Chapel and St Peter's Church, Heysham, Lancashire 1977–1978. *Antiquaries Journal* LXXIV, 54–134.

Redknap, M. (2004). Viking-age settlement in Wales and the Evidence form Llanbedrgoch. In Hines, J., Lane, A., and Redknap, M. (Eds.), *Land Sea and Home: Settlement in the Viking Period*, Society for Medieval Archaeology Mnography 20, Maney, Leeds, pp. 139–175.

Richardson, C. (1996). A find of Viking period silver brooches and fragments from Flusco, Newbiggin, Cumbria. *Transactions of the Cumberland and Westmorland Antiquarian and Archaeological Society* 96, 35–44.

Rooke, H. (1792). Druidical and other remains in Cumberland. *Archaeologia* 10, 105–113.

Stocker, D. (2000). Monuments and merchants: irregularities in the distribution of stone sculpture in Lincolnshire and Yorkshire in the 10th century. In D.M. Hadley and J.D. Richards, Eds., *Cultures in Contact: Scandinavian Settlement in England in the Ninth and Tenth Centuries.* Brepols, Turnhout, 179–212.

Thacker A.T. (1987). Anglo-Saxon Cheshire. In B.E. Harris and A.T. Thacker, Eds., *The Victoria History of the County of Chester,* Vol. 1. Institute of Historical Research, London, pp. 237–292.

Thompson-Watkin, W. (1883). *Roman Lancashire* (reprinted in 1969 by S.R. Publishers), Thomas Brakell, Liverpool.

Wainwright, F.T. (1948). Ingimund's invasion. *English Historical Review* 63, 145–169.

Ward S.W. (1994). *Excavations at Chester: Saxon Occupation within the Legionary Fortress, Sites Investigated 1963–1981.* Chester Archaeological Service Monograph 7, Chester.

Whitelock, D. (1979). *English Historical Documents,* Vol. 1: 500–1042. Eyre Methuen, London.

Wilson, D.M. (1984). *Anglo-Saxon Art from the Seventh Century to the Norman Conquest.* British Museum Press, London.

Youngs, S. (2002). Cumbria, Arnside. *Medieval Archaeology* 46, 129–130.

Youngs, S. and Herepath, N. (2001). Cumbria, Arnside. *Medieval Archaeology* 45, 237–238.

3

Speaking Like a Viking: Language and Cultural Interaction in the Irish Sea Region

Judith Jesch

CONTENTS

ABSTRACT   This chapter is an attempt to reconstruct some of the linguistic and cultural conditions in which the Viking Age genetic input of Scandinavians into the population of north-west England took place. Starting from the observation that Y-chromosome studies cannot reveal much about the role of women in this process, the evidence of language, place-names, runic inscriptions, and carved stones is adduced to produce a picture in which women were essential elements in the creation and maintenance of the Norse-speaking diasporic communities of the Irish Sea region in the Viking Age.

Introduction

In recent years, a range of both scientific and popular publications have explored the extent to which genetics can reveal the input of Scandinavian settlers into the population of north-west England (Bowden et al. 2008; Harding et al. 2010). These genetic results seem exciting but what exactly do they reveal? The limitations of modern population genetics in explaining the past are well known. These studies, for instance, focus on the Y-chromosome, so can only tell us about male input into the population. Even within that restriction, the results tell only a part of the story.

The Y-chromosome represents the direct paternal line, and over a thousand years, this is a very small part of the genetic ancestry of any one individual. Moreover, genetic proportions are not necessarily the same in modern populations as in past ones—some lines die out, others are successful, and some emigrate. And the genetic results can be interpreted in various ways. In theory, they are not incompatible with a situation in which a lot of Vikings passed through the north-west of England and impregnated a lot of women and then left again. If this were the case, the Viking contribution to the heritage of the north-west would be purely genetic, and we would have nothing else to study. However, it was clearly not that simple and there is evidence that the genetic input occurred in the context of other sorts of human interactions. This chapter will therefore explore the linguistic and cultural contexts within which any genetic interactions took place.

The focus of this chapter is on communities of people around the Irish Sea in the Viking Age.* To flesh out the skeletal picture provided by the genetic evidence, the following questions about these communities of people need to be considered:

- What language or languages did they speak?
- What was their family life like?
- What religion did they profess?

These are all relevant to the interpretation of the genetic evidence, since they raise questions that the genetic evidence alone cannot answer.

Communities and Diaspora

The study of such communities requires first a few thoughts on the terms 'migration' and 'diaspora' (Clifford 1999; Cohen 2008; Jesch 2008a). Human history has always been about migration; 18,000 years ago no one lived in north-west England because it was covered in ice. After the ice receded, people began migrating northward. Immigration and emigration happen throughout history, more intensely in some periods than others.

We live in an Age of Migration now, and the 10th century was also such a time in the region around the Irish Sea, the M6 motorway of its time, absolutely clogged with traffic. Just as on the M6, much of this traffic was commercial and involved trading activities. But some of it also involved people upping sticks and moving to new places. This is *migration*—the simple transplantation of people from the place in which they were born to another place where they settle and remain (some of them may even go on to somewhere else). Migration may involve staying within the same region or country where the language, culture, social and economic customs, and religion are the same or very similar, although such moves tend not to be thought of as migrations. Migration is usually understood to mean that people move somewhere with a different language, culture, social and economic customs, and religion. And in a historical context, the term 'migration' is used only when large numbers of people make the same move.

Diaspora, on the other hand, relates to the processes and results of migration and how migrants think and feel about their situation. Did they migrate in a group? Did they migrate to a place where there were other migrants from their home? Did they take their own social and cultural customs with them or did they adopt new ones? Did they give their children traditional names and encourage them to speak the old language as well as the new one? Did they assimilate into the culture of their new homes, and if so how many generations did that take? (How indeed does one define *assimilation*?) Did they maintain connections with their homeland? If not, did they nevertheless have a sense of where they had come from and a memory of how things were there? And were they in touch with other migrants from the same homeland who migrated somewhere else entirely? In other words, diaspora is about the migrant's sense of connectedness:

- To the homeland
- To other migrants from the homeland
- To other regions with migrants from the same homeland
- To the new home

After the actual migrations had taken place, there remained webs and networks of connections among all of these groups, and that is diaspora—an ongoing connectedness arising from a migrational event or events. Nowadays it is possible to study these matters by asking people about their experiences, or, in the

---

* It will not be possible to give consideration to the complex matter of the Viking Age in Ireland, but see Sheehan and Ó Corráin 2010.

case of people who themselves have migrated, by introspection. However, for the Viking Age, we have no facts and figures about who migrated, or when and where, and it is no longer possible to ask the people who did so about their experiences or their feelings of connectedness. The answers to these questions have to be teased out of limited evidence.

Genetic studies, despite their limitations, can helpfully point the way. Although they do not tell us much about the people living in the north-west of England and the Irish Sea region in the Viking Age, they suggest interesting questions. The most glaring question that remains unanswered by the Y-chromosome studies concerns the role of women. Did the Vikings simply pass through and leave a lot of babies behind or was there a migration of families including women and children?

Language

Genetic studies also raise the question of what else might be inherited along with genes. An important aspect of the individual as a human being is the use of language (Joseph 2004). While not as firmly fixed in the physical body as genes are, language is nevertheless closely associated with the body, developing along with it in infancy and childhood, within the social context of family and community. In normal circumstances, language is acquired at a very young age from the people surrounding an individual, especially immediate family members. Most languages have a concept of 'the mother tongue' since an infant has the closest contact with his or her mother at the stage of life when language is acquired. Because of this close contact, infants are most likely to learn their mothers' language first. For many people, language development ends when they have acquired their language. They speak that one language for the rest of their lives, particularly if it also the language of their fathers, other family members, and their communities.

It is certainly possible to learn other languages, and some people can become equally fluent in two or more languages if they use them regularly; others make a complete switch from one language to another, while yet other people end up in a situation where they are not totally at home in two or more languages. A lot depends on the circumstances and the age at which the languages are learned, the situations in which the individual needs to use one or the other, and the extent to which he or she has contact with other speakers of that language. We do not know much about these circumstances in the Viking Age, but we need to investigate the evidence with these questions in mind.

If it transpires that the Old Norse language as spoken by the Vikings was in fact used in north-west England, this immediately puts into question the hypothesis of marauding Vikings having their wicked way with the virtuous ladies of the region and then leaving them in the lurch. There is in fact a lot of evidence (mainly from place-names; see below) for the Old Norse language having been spoken in this region in the 10th century and for some time afterwards. If the language was spoken to the extent suggested by this evidence, then that is very good evidence for the presence of women speaking the language. People can, of course, learn more than one language, as already noted. So, evidence that the Old Norse language was spoken by women does not necessarily imply that the women came from the Viking homelands such as Norway. It is possible to imagine that the marauding Vikings who came and impregnated the ladies of the region then made honest women of them and settled down with them in places later known as Ormskirk, Croxteth, or Skelmersdale. Having done so, they taught those women to speak Old Norse and persuaded them to speak it with their children. This is, however, more likely in some situations than others. For example, if a hypothetical Viking kidnapped and enslaved a girl in Ireland and then took her to Skelmersdale, along with a load of other Vikings who had done the same thing, it is perfectly possible and even likely that she would have had to learn Old Norse as well as her native Irish to function in this group. A lot may have depended on whether she remained his slave. It was quite common for men in the Viking Age to have children with their slave women but less common for them to marry the women and set up a house and family with them.

This situation is illuminated by a well-known story from an Icelandic saga (Kunz 2000, pp. 286–290). It is set in Iceland in the 10th century and involves a powerful man who buys a beautiful slave girl and takes her home to Iceland. She appears to be mute and refuses to speak to anyone. When the man brings his new slave home to Iceland, his wife sarcastically points out that he must have spoken to her at some point since she seems to be with child. The slave and her son are given a small place of their own and the chieftain finds out that she can actually speak only to him when one day he overhears her speaking to her small son in Irish. It turns out that the slave is an Irish princess and her son becomes a heroic figure in the saga. When he eventually gets to Ireland, he is able to speak to his family by using the linguistic skills passed on by his mother. The saga never actually tells us whether she learned Old Norse but we must assume she did because she later married someone else in Iceland where she ended her days.

Although this saga was written several centuries after the Viking Age, the picture rings true. If such a situation had happened in north-west England, it could have resulted in a Norse-speaking community although if it happened on a large scale we would rather expect a bilingual community. There is some evidence that this is indeed what happened. A number of place-names show just this linguistic inter-action between Norse and Irish speakers. However, these are more frequent in Cumbria than further south and even there the examples do not suggest that much Irish was spoken.

No names in the north-west are entirely Irish in the way that there are names that are entirely Old Norse. The most common Old Irish word in place-names is *erg*, a word whose meaning is not entirely certain but which probably refers to a hut or building used in summer pasturing arrangements, or a 'shieling' as it is often translated. This word seems to have been borrowed into Old Norse and is thus not genuine evidence for the speaking of Irish. The Viking settlers learned it somewhere in the Irish Sea region and found it useful even though they already had their own word for summer pasturing arrange-ments (*setr*), found in names like Swainshead, Hawkshead, or Arnside.* So, compounding of the *erg* ele-ment with an Old Norse personal name (as in Anglezark, Grimsargh, or Sholver) is really only evidence for Old Norse speech. Further evidence for the naturalisation of this word in Old Norse is that we find it appearing with Old Norse grammar (i.e., the dative plural form) in Arkholme.

Occasionally we find the *erg* word compounded with an Irish name, as in Goosnargh. This might in theory be taken as evidence for Irish speech since both elements are Irish. However, personal names are not very good indicators of ethnicity; they tend to follow fashion and even cross linguistic boundaries. The more important element in a place-name is always the second one. The Becconsall name, which consists of an Irish personal name and the Old Norse element *haugr* ('mound') is evidence of Norse linguistic usage, not Irish. The mound may have belonged to a person with an Irish name who may or may not have spoken Old Norse, Old Irish or both, but the people who gave the mound its name used the Old Norse language to identify it. The chap called Beccan may have been the only Irish person (or just a person with an Irish name) in a Norse-speaking community. Similarly, even Goosnargh could be a name given in a Norse-speaking community. Thus, the place-name evidence in England does not suggest a large Irish-speaking population accompanying the Vikings although it reveals traces of that contact before the Vikings got here.

When the Vikings arrived, north-west England was already populated by people speaking another lan-guage, i.e., Old English. Here, too, are interesting questions. If a lot of Vikings arrived and settled down with local women, what would the linguistic situation be? Did they teach their wives to speak Old Norse and form a Norse-speaking community? This is surely less likely than in the case of Irish wives, given that the English women would be on their home territory and still plugged into their family networks and local cultures. We know that Viking settlements tended to be in areas that were sparsely or not inhabited by the locals but even so any such communities would not be that far away from English-speaking ones. It is difficult to believe in Norse-speaking communities in this context if the women were not native speakers of the language, although bilingual communities do seem possible. For there to have been any extensive use of the Old Norse language, there must have been at least some women whose first and most

---

* Interpretations of place-names in this chapter are derived from the following sources: Armstrong et al. 1950–1952, Dodgson 1972, Ekwall 1922, Fellows-Jensen 1985, Mills 1976, Smith 1967, and Whaley 2006. It should be noted that the *Survey of English Place-Names* has not yet covered Lancashire, and that some of the interpretations in this county are therefore more provisional than usual.

natural language was Old Norse. Otherwise, the men would have taken the line of least resistance and learned English pretty quickly. So language tells us something that genetics does not.

Whereas the place-name evidence for Irish-Norse bilingualism is not strong, it is much stronger for Old English and Old Norse (Townend 2002). There are a number of reasons for this. One is that Old English and Old Norse were fairly closely related languages and it was quite easy for anyone who knew one to learn the other. The main reason, however, comes from the processes of migration and diaspora. North-west England had a well-established English-speaking Christian population. The area may have been fairly sparsely populated at the time. One possible explanation for the Viking settlements is that there was plenty of room for them in amongst the English settlements, though possibly on slightly inferior land since the English settlements would have been on the better land. The establishment of Norse-speaking communities who would have given their localities names in Old Norse would have required substantial migration of Norse speakers, including women and children, who maintained their language and culture.*

Another reason for assuming Anglo-Norse linguistic contact is the question of loan-words. Just as the Vikings borrowed the Irish *erg*, a lot of Old Norse words entered the English language. Once they become part of the English language, they are no longer direct evidence for Viking Age speech communities because they spread through the language, and many such words are still widely used in English (Barber 2000, pp. 130–134). Many words of Old Norse origin occur in place-names, but the names may have been given by English speakers who spoke a dialect in which these words had been assimilated. Examples include words like *rake* (narrow path) and *carr* (marsh or bog), both common in the north-west (Harding 2007). Nevertheless, the number of such words that entered the English language does suggest the extent of previous linguistic contact that in turn suggests there were relatively large numbers of people speaking Old Norse, who lived in communities where Old Norse was the first language.

Place-Names

One interesting aspect of Scandinavian place-names in the north-west, compared to other parts of England, is how many of those names describe the landscape rather than referring to settlements or are what onomasts call 'topographical' rather than 'habitative.' This does suggest that the landscape was quite thinly inhabited and that the incoming Vikings did not have much contact with the natives to find out what particular mountains, rivers, beaches, or woods were called, so they gave their own names to the landscape around them and to their farms and villages. This is particularly noticeable in Cumbria, but there are also plenty of examples in Lancashire and the Wirral. There are even river names in Old Norse, which is fairly unusual as river-names tend to be the oldest in most landscapes, handed on from one linguistic group to another. Thus many becks like Artlebeck and the rivers Winster and Greeta are entirely Old Norse in form. Other names describe the landscape. Aintree is a place with a solitary tree. Blawith is a dark wood and Blowick a dark bay, while Tranmere is a sandbank frequented by cranes. All these names derive from personal observation and familiarity with the landscape.

Some of the most interesting names are those that reveal some of the detail of the Scandinavian language. It has already been noted that Arkholme comes from a dative plural form, providing insight into the grammatical structure of Old Norse. Another example is Litherland. This name consists of two elements that are identical or nearly identical in Old English and Old Norse: *hlíð* slope and *land* land, thus a piece of 'land' on or near a 'slope'. What makes the name Norse rather than English is the *er* in the middle of the modern form, which is a remnant of an Old Norse genitive singular form, i.e., 'the land of the slope' (the Old English word would have had no ending in this form). In this way, grammar reveals history, as the name was given by someone who spoke grammatically correct Old Norse.

---

* Abrams and Parsons (2004) make a strong case for the value of place-names as evidence for substantial communities of speakers of Old Norse although their work is based mainly on evidence from eastern England. See also Jesch (2008b).

Even names that have an English origins can show the influence of Scandinavian speech. The sound combination *sk* in a word or a name is often an indication of Norse origin, so that 'fish' is *fisc* (pronounced 'fish') in Old English but *fiskr* in Old Norse. This is the kind of word a Viking immigrant could use and make himself understood when buying or selling fish at an English market, for instance. Where there were lots of Norse-speakers around, names with a *sh* sound in Old English could develop a Norse pronunciation as in the example of Fishwick, early spellings of which suggest it was once Scandinavianised and pronounced something like Fiskvik. Similarly, the Old English word *stan* ('stone') has a very close equivalent in the Old Norse *steinn* with the same meaning. Hence, a place called Stainton is simply a place called Stanton pronounced with a Scandinavian accent.

A final indication of Viking-speaking communities in the north-west comes in the two places called Thingwall on both sides of the Mersey. This name means 'assembly field' and refers to a common and important social institution in the Viking world: the *thing* or regular assembly where people met to make and enforce laws, bring lawsuits against enemies or neighbours, and generally to gossip, tell stories, and recite poetry. It was also a good place to meet potential marriage partners in sparsely populated areas. All these functions are well evidenced in the Icelandic sagas that show the operation of the *thing* in Iceland and to some extent in Norway. There is no direct evidence as to how such assemblies operated in the Irish Sea region but they are likely to have been similar. Because of the nature of the institution, the Old Norse language was very much used and maintained in the diasporic community.

Language, Sculpture, and Stories

A version of the same place-name is found on the Isle of Man in Tynwald where the legal, political, and cultural functions are still evident. The Isle of Man is an interesting example in contrast with north-west England (Fell et al. 1983; Wilson 2008). It, too, had a major influx of Scandinavian settlers, but into a community where a Celtic language rather than English was spoken. There is contemporary evidence for the use of the Old Norse language in the form of runic memorials. The memorials also provide important evidence for the maintenance of cultural traditions in the diasporic community.

The Isle of Man has some 30 carved stones with crosses, both abstract and figural decorations, and runic inscriptions. It also has a number of other stones with crosses and decoration but without runes. These date on the whole to the 10th century. The runic inscriptions are very important for our understanding of the use of Old Norse then as they represent a cultural phenomenon imported from Norway. Runes constitute an alphabet used for inscriptions but not books. Runes were the only forms of literacy in Scandinavia during the Viking Age. Inscriptions in this alphabet are very important contemporary records of the language of that period. There are many interesting aspects of the Manx inscriptions but the focus here will be on what they reveal about language and gender.

The inscriptions are memorial texts on monuments erected by family members to commemorate their dead. Both men and women are commemorated, and the inscriptions mention one or more members of the family commemorating them. Both the runes and the language of the inscriptions are Scandinavian, but they show signs of contact with Celtic. First, the forms of the monuments (most of which are crosses) and some of their decoration are obviously insular. Second, a number of the names in the inscriptions are Celtic. Third, some of the inscriptions contain grammatical mistakes that suggest that some of the users of Old Norse either had another first language or that the Norse language in the Isle of Man had developed certain peculiarities that differentiated it from the language of the Norwegian homeland— peculiarities that may have arisen in a bilingual context.

The range of possibilities is revealed in the contrast between two inscriptions, one from Andreas and one from Kirk Michael.* In the inscription on Andreas II ('Sandulf the Black raised this cross in memory of his wife Arinbjorg'), both personal names are Old Norse and there are no grammatical errors. This

---

* Translations of these two inscriptions are my own. The inscriptions have not yet been given a full scholarly edition, but provisional texts and linguistic discussion can be found in Page 1983. A number of the monuments are depicted in Fell et al. (1983) and Wilson (2008).

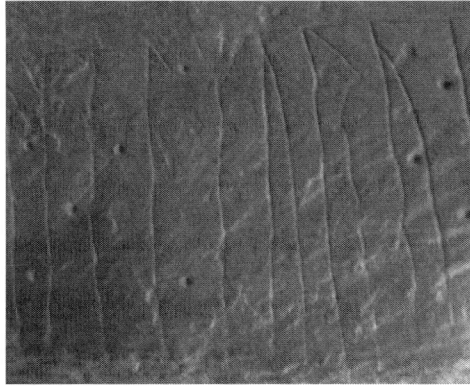

FIGURE 3.1 Part of the runic inscription on the Kirk Michael III cross, Isle of Man (MM 130) showing the name 'Malmury.' (Copyright Judith Jesch.)

suggests an emigrant family of Norse speakers and probable Norwegian origin on both sides. The Kirk Michael III inscription, by contrast, is linguistically more complex ('Mallymkun [?] raised this cross in memory of Malmury [?], his foster mother (?), daughter of Dufgal [and] the wife to whom Adisl was married. It is better to leave a good foster son than a bad son.'); see Figure 3.1.

Despite some uncertainties in reading this inscription, it is clear that three of the four names are Celtic, yet the text is still in the Old Norse language, albeit with some grammatical peculiarities, and in Scandinavian runes. The people mentioned are a woman with a Celtic name, her father with a Celtic name, her husband with a Norse name, and her foster son with a Celtic name. If it were the husband commemorating his wife, as in the example from Andreas, then it would be no surprise that the inscription is in Old Norse, as he would have decided this matter. But the commissioner of the monument in this case is the foster son with a Celtic name. Did he choose Norse in honour of the language spoken by his adoptive family or because it was the only option for this kind of monument? Did he even speak or understand Old Norse? Are at least some of these people in fact Norse speakers who adopted new fashionable names from their Celtic environment?

The latter seems less likely because both the names and the bad grammar of the inscription are more suggestive of a mixed family. Despite this, the family chose to continue cultural traditions from the Scandinavian homeland. The social custom of fostering someone else's child was very strong in the Viking world, and of course both the erection of a stone monument and the use of runes conform to long-standing Scandinavian traditions. Another cultural tradition from the homelands maintained by the diasporic community on the Isle of Man is that of telling old stories of gods and heroes. Several of the carved stones of the island are decorated with motifs that correspond to legends and stories that were written down much later, in medieval Iceland. One of these is about a hero known as Sigurd Fafnisbani.[*] The story involves a great treasure guarded by a serpent, Fafnir. The serpent's brother is a blacksmith who feels that the treasure should be his. He persuades the young hero Sigurd to try to kill Fafnir on his behalf, and provides Sigurd with a sword to kill Fafnir. Sigurd does indeed kill the serpent and get the treasure, without at first realising that the smith hopes to kill him and claim the treasure.

Luckily, in his moment of victory, Sigurd roasts the heart of the serpent on a fire. Before eating it, he tests it with his thumb to see if it is done, burns his thumb slightly and therefore licks it. As a result of tasting the serpent's blood, he can understand the speech of birds, and hears from two birds in a tree that the smith intends to kill him. Sigurd acts first, kills the smith, and takes the treasure away on the back of his faithful horse Grani. This story is widespread in the Viking world and Wagner fans will recognise a version of it from his Ring cycle of operas. The story was very popular in both England and the Isle of Man in the Viking Age, and allusions to it are found on a large number of sculptures (Margeson 1980).

---

[*] The story is succinctly told in Faulkes (1987, pp. 99–102).

FIGURE 3.2 Face C of the Halton 1 cross, Lancashire, showing scenes from the story of Sigurd Fafnisbani. (Photograph courtesy of Ross Trench-Jellicoe.)

FIGURE 3.3 Face D of the Halton 1 cross, Lancashire, showing scenes from the story of Sigurd Fafnisbani. (Photograph courtesy of Ross Trench-Jellicoe.)

One of these sculptures is found at Halton, near Lancaster (Bailey 2010, pp. 177–183). Two panels of this cross are shown in Figures 3.2 and 3.3. The panels show clearly identifiable scenes from this story: the smith forging the sword for Sigurd, with all his tools around him, the serpent looking a little bit dead, Sigurd roasting the heart of the dragon on a spit and putting his thumb in his mouth, above him two birds sitting on some branches, and finally Sigurd's horse Grani. This sculpture actually includes all the main elements of the story as we know it from Old Icelandic literature, demonstrating the diasporic connections between the homeland and two areas of migration, the Irish Sea region and Iceland, across several centuries.

Diaspora in Action

These sculptures represent the diaspora in action. Both in the Isle of Man and in north-west England, they show communities remembering their cultural traditions from the homeland. But, these communities lived in very different cultural and linguistic circumstances—alongside Celtic-speaking communities in the Isle of Man and alongside English-speaking communities in England—and they used these cultural traditions in different ways. In the Isle of Man they developed a distinct local custom of erecting memorial stones for the dead, often with runic inscriptions. In north-west England they also developed local customs of sculpture, though generally without the use of runic inscriptions. The local custom in England adopted local cultural traditions; although the Scandinavians were familiar with stone monuments, they did not have the custom of figural and decorative carving in stone; that was something they learned in England. Moreover, like the Halton cross, most of the monuments they carved are very clearly

FIGURE 3.4 Detail of the Gosforth cross, Cumbria, showing the Crucifixion. (Copyright Judith Jesch.)

Christian, even when they show images from Norse myth and legend. The most famous example is the great cross at Gosforth in Cumbria (Bailey and Cramp 1988, pp. 100–104); see Figure 3.4.

This is what diaspora is about. The migrants and their descendants are still in touch with their cultural traditions and their language, both brought from the homeland, but they also develop new cultural traditions, borrowing elements from their new neighbours. Such sculptures cannot be described as Viking in any meaningful sense, except insofar as they are the result of the Viking diaspora. As well as borrowing cultural traditions, the migrants and their descendants may also have borrowed the religion of their neighbours. The Vikings were not Christians when they left Norway, but by the time they were commissioning the runestones of Man or the Halton cross they clearly were. The way this happened is still not fully understood, but somewhere along the way they decided to start worshipping Christ. One possibility is that women played a role. If some of the Vikings who settled in the Irish Sea region married and established families with local women, who were of course Christian, it is very likely that those women would have encouraged them very strongly to convert. It is also possible that women of Norse origin who came to the Irish Sea region with their Viking husbands, may have been more receptive to the Christian message and encouraged the conversion of the whole community. There is some evidence that this happened back in Scandinavia and in other Viking colonies such as Greenland. If women were the agents of change in religion, it is also true that they often had the role of maintaining cultural traditions such as telling the old stories of heroes like Sigurd Fafnisbani. By doing so, they also maintained the language and passed it on to the next generation.

Although women are not so visible in either historical records or the archaeological evidence, they are in fact the key to understanding what really happened in the Viking Age. What the genetic evidence reveals is that there was a lot of Viking sperm about. But the evidence of language and culture in the form of place-names, inscriptions, and sculpture provides a much fuller picture of the community. It shows women maintaining the language and cultural traditions of the homeland while also being receptive to new ideas about religion, and new expressions of identity, both borrowed and transformed from their new neighbours. This is what diaspora is about—not merely a transplantation of people from one country to another, but a whole new culture formed by combining the memories of the old country with the best of the new country.

ACKNOWLEDGMENTS

An earlier version of this chapter was given as the first Ruth Vose Memorial Lecture in Ormskirk on 21 January 2009. I am grateful to the West Lancashire Heritage Association (and particularly Patrick Waite and Jim Vose) for the invitation, for being an enthusiastic and attentive audience, and for their splendid hospitality on that occasion. I am also grateful to the editors of this volume for inviting me to publish this chapter to complement presentations given at the Chester Viking Conference on 20 November 2010. Thanks are also due to Dr. Derek Craig of Durham University, for his assistance with photographs.

REFERENCES

Abrams, L. and Parsons, D.N. (2004). Place-names and the history of Scandinavian settlement in England. In *Land, Sea and Home*, J. Hines, A. Lane, and M. Redknap, Eds. Maney, Leeds, UK, pp. 379–431.

Armstrong, A.M., Mawer, A., Stenton, F.M. et al. (1950–1952). *The Place-Names of Cumberland I–III*. Cambridge University Press, Cambridge, UK.

Bailey, R.N. (2010). *Corpus of Anglo-Saxon Stone Sculpture IX. Cheshire and Lancashire*. Oxford, Oxford University Press.

Bailey, R.N. and Cramp, R. (1988). *The British Academy Corpus of Anglo-Saxon Stone Sculpture II. Cumberland, Westmorland and Lancashire North-of-the-Sands*. Oxford, Oxford University Press.

Barber, C. (2000). *The English Language: A Historical Introduction*. Cambridge, Cambridge University Press.

Bowden, G.R., Balaresque, P. King, T.E. et al. (2008). Excavating past population structures by surname-based sampling: The genetic legacy of the Vikings in north-west England. *Molecular Biology and Evolution*, 25, 301–309.

Clifford, J. (1999). Diasporas. *Cultural Anthropology*, 9, 302–338.

Cohen, R. (2008). *Global Diasporas: An Introduction*. London, Routledge.

Dodgson, J. McN. (1972). *The Place-Names of Cheshire IV*. Cambridge, Cambridge University Press.

Ekwall, E. (1922). *The Place-Names of Lancashire*. Manchester, Chetham Society.

Faulkes, A., Trans. (1987). *Snorri Sturluson Edda*. London, Dent.

Fell, C., Foote, P., Graham-Campbell, J. et al. (1983). *The Viking Age in the Isle of Man*. London, Viking Society for Northern Research.

Fellows-Jensen, G. (1985). *Scandinavian Settlement Names in the North-West*. Copenhagen, C.A. Reitzels Forlag.

Griffiths, D. (2010). *Vikings of the Irish Sea*. Stroud, History Press.

Harding, S. (2007). The Wirral *Carrs* and *Holms*. *Journal of the English Place-Name Society*, 39, 45–57.

Harding, S.E., Jobling, M. and King, T. (2010). *Viking DNA: The Wirral and West Lancashire Project*. Birkenhead, Countyvise.

Jesch, J. (2008a). Myth and cultural memory in the Viking diaspora. *Viking and Medieval Scandinavia*, 4, 221–226.

Jesch, J. (2008b). Scandinavian women's names in English place-names. In *A Commodity of Good Names*, O.J. Padel and D.N. Parsons, Eds. Donington, Shaun Tyas, pp. 154–162.

Joseph, J. (2004). *Language and Identity: National, Ethnic, Religious*. London, Palgrave.

Kunz, K., Trans. (2000). The saga of the people of Laxardal. In *The Sagas of Icelanders: A Selection*, Örnólfur Thorsson, Ed. London, Penguin Books, pp. 270–421.

Margeson, S. (1980). The Völsung legend in medieval art. In *Medieval Iconography and Narrative*, F.G. Andersen, E. Nyholm, M. Powell et al., Eds. Odense, Odense University Press, pp. 183–211. Odense University Press, Odense, Denmark.

Mills, D. (1976). *The Place-Names of Lancashire*. London, Batsford.

Page, R.I. (1983). The Manx rune stones. In Fell et al., Eds. 133–146.

Sheehan, J. and Corráin, D.Ó., Eds. (2010). *The Viking Age: Ireland and the West*. Dublin, Four Courts Press.

Smith, A.H. (1967). *The Place-Names of Westmorland I–II*. Cambridge, Cambridge University Press.

Townend, M. (2002). *Language and History in Viking Age England*. Turnhout. Brepols.

Whaley, D. (2006). *A Dictionary of Lake District Place-Names*. Nottingham, English Place-Name Society.

Wilson, D.M. (2008). *The Vikings in the Isle of Man*. Aarhus, Aarhus University Press.

4

Viking Age Women

Christina Lee

CONTENTS

ABSTRACT   Women were involved in many of the Viking migrations and this chapter considers what types of evidence we can use to make them visible in the regions of Ireland and the British Isles. Textile production is especially useful when looking at female lives and this chapter will discuss the production and manufacture of some textiles in the Irish Sea region as well as influences on dress in the Viking diaspora. Place-names too render some convincing evidence about the longevity of Old Norse names in the Irish Sea region and they can sometimes even give us a glimpse of the people who were part of this community.

Introduction

How do you identify a Viking woman? At first glance there seems to be a pretty obvious answer, but it is not so straightforward. Are you a Viking woman when you are ethnically Scandinavian, or could you also be a Viking woman when you married into the culture? The question has become ever more complex since the arrival of DNA analysis that has shown that many modern Icelandic women have Irish ancestors (Helgason et al. 2001).

It has always been claimed that there are few loan words of Irish–Gaelic origin in the Icelandic language, which has been regarded as evidence for poor integration of the two societies, but ongoing work by Rod McDonald (forthcoming) shows a considerable amount of vocabulary shared across the divide. This is not surprising, since we have groups who appear to be culturally Norse, but are described as 'Gaelic speaking foreigners' (Kruse and Jennings 2009). Would these 'foreigners' have included women as well? Or did the women all stay at home?

One way of looking for women in the Viking diaspora is to search for their handiworks. Textile work was female work in the early Middle Ages. We know this from taxation paid in cloth by women and text descriptions of women weaving and embroidering. In this chapter, I want to examine where we might find these women in the north-west England and wider Irish Sea context.

Textiles and Viking Age Women

Textile tools are ubiquitous in a Viking context. The finding of a spindle whorl at the furthest westerly expansion of the Vikings that we know of in L'Anse aux Meadows, Newfoundland, Canada and the

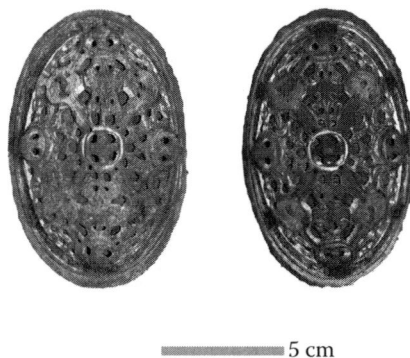

5 cm

FIGURE 4.1 Gilded copper alloy composite oval brooches from the Viking cemetery at Cumwhitton, Cumbria. (*Source:* Oxford Archaeology North. With permission.)

recent excavations by Patricia Sutherland have shown that European-style cordage made from animals such as hares and foxes can be found as far as Baffin Island (2009). Such evidence suggests that Viking expeditions included at least a few textile workers among the explorers. Since the spinning of very fine hairs such as those of hares requires a very fine skill level, Sutherland's work suggests that experienced female craftworkers participated in Viking explorations.

Textile work in the Viking Age was taught presumably from mother or foster mother to daughter, so the techniques were passed along generations. This transfer of skill should allow us to research differences in techniques, such as the spinning of the yarn. The widely known depiction of the quintessentially Viking dress consisting of a pinafore pinned with oval brooches on the shoulder, is largely based on costumes found at the Swedish cemetery of Birka, reconstructed by Swedish archaeologist Agnes Geijer (1938). However, this analysis is based on only 5% of all recovered fragments from the cemetery.

The work of subsequent textile historians such as Inga Hägg (1991) and Lotte Hedeager, and Eva Andersson-Strand (2011) has shown a much larger variety of both fabrics and fasteners in the homelands. Thus, the ubiquitous Viking woman's dress fastened with oval brooches (Figure 4.1) was already becoming unfashionable in Denmark when many of the Scandinavian settlers to England left their homelands. Far fewer oval brooches were recovered from Denmark than were found in Norway or Sweden. Most of these brooches are from a 9th century context and were found in the eastern part of the country that is closest to Sweden.

At the Birka cemetery, however, almost every grave contained oval brooches. The remains of fabric on the backs of the brooches can tell us a little about the social status afforded to the dead. The fabric ranged from poor coarse woollen material to fine silk imported from the Orient. Costume in the past has been taken as a marker of ethnic origin. However, the correlation of dress and ethnicity is very tentative and should be regarded with a healthy dose of skepticism. Changes in fashion indicate that even in the Viking Age the 'one size fits them all' garment is a myth.

And this leads to other important questions: When did changes happen? Who adopted them? Were certain types of clothes worn by every woman? Could such fabrics be regarded as markers of a Norse identity? If this is the case, how do we know that the dress of the homelands and the dress of the Viking diaspora served the same purpose? Not all of these questions can be answered, but I think they are important to ask.

Most recreations of Norse dress have been based mainly on evidence found in graves (see Figure 4.2), posing the question whether funeral wear was the same as garments worn by living populations. A cautionary tale may be the travel report by an Arabic envoy named Ibn Fadhlan who met a group of Rus (taken to be Viking Age Scandinavians or a group of mixed Viking and Khazar people) on the Volga. He observed the funeral of a wealthy man and noted that new clothes were made for the man in preparation for the funeral. How do we know, then, if people were not dressed specifically for the grave?

Most of the grave dress found in the British Isles came from rural environments and which may have been different from urban dress. Grave goods have symbolism and we now think that they are meant to

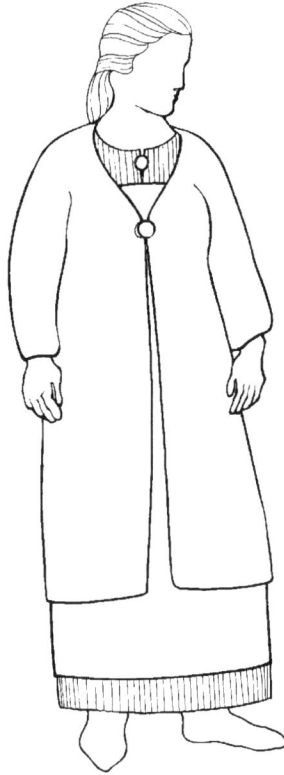

FIGURE 4.2 Female costume documented from Hedeby, Schleswig Holstein; drawing by Tina Borstam in Andersson-Strand 2003 (after Hägg 1991). (*Source:* Eva Andersson-Strand, Copenhagen Textile Research Project, http://ctr.hum. ku.dk/. With permission.)

tell a story about the dead—position in society, wealth, and perhaps age (Williams 2006). If this is the case, textiles are part of this narrative, and we should and must consider them as having special meanings.

Currently, archaeological publications commonly list textile finds in appendices, divorced from catalogues of grave goods. If the residues on the backs of brooches are supposed to represent the funerary attire of the dead, they should be included in grave inventories. Dress historians, however, must be careful since some of the imprints found on the backs of the brooches were not parts of the clothing worn by the deceased and were made during the manufacturing process. Clay-impregnated cloth was used to separate a bronze cast from its mould. When the hot metal was poured into the mould, the textile disintegrated and left an imprint. Unless a textile historian participated in the identification, such poor quality cloth could easily be taken as the clothing of the dead. Also, not all of the cloths used were from cheap rags. A very fine herringbone twill was found on the back of a brooch at Chaipaval on Harris in the Outer Hebrides (Graham-Campbell 1975).

The spread of oval brooches around the Irish Sea region indicates that at least in death high-ranking women continued to be buried in the fashion of their ancestors. Grave finds in Scotland such as at Westness in the Orkneys, Cumwhitton in Cumbria, and Finglas in County Dublin all contained pairs of oval brooches. However, in areas of eastern and midland England that were also Norse settled and commonly called the Danelaw, people seem to have preferred Irish-influenced ringed pins or penannular brooches rather than oval brooches.

Textile production occupied much of a woman's time. The twelve ells of yarn that Gudrun, the heroine of the *Laxdæla* saga, produces in one morning may be an exaggeration, but textile production required a lot of time and energy. For this reason, some textiles may not have been produced at home.

FIGURE 4.3 Wool combing. Drawing by Annika Jeppsson in Andersson-Strand 2003. (Source: Eva Andersson-Strand, Copenhagen Textile Research Project, http://ctr.hum.ku.dk/. With permission.)

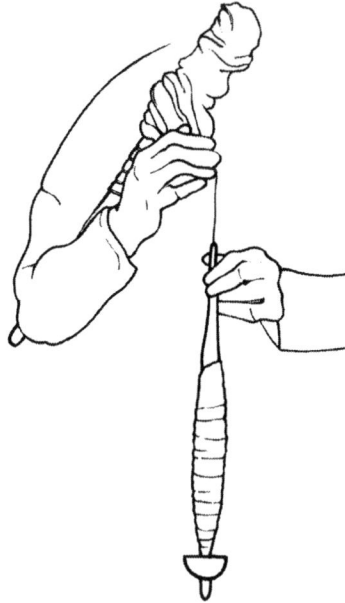

FIGURE 4.4 Spinning with a drop spindle. Drawing Tina Borstam in Andersson-Strand 2003. (Source: Eva Andersson-Strand, Copenhagen Textile Research Project, http://ctr.hum.ku.dk/. With permission.)

Reconstructions of Viking Age dress have shown that it takes 2.5 kg of wool to produce the most basic female dress and 3.5 kg to make a man's costume (Andersson-Strand 2011, p. 2). One sheep produces around 2 kilos of wool and such an amount must have been difficult to obtain in the Viking Age. Sheep were not shorn routinely until the 12th century in Scandinavia. Wool was also obtained by 'rooing'—the upper coat of wool was hand-plucked from sheep when they were moulting, leaving the new hair intact.

This may have led to preferences for different breeds of animals such as long-haired sheep. Analysis of wool has shown that the fleece types and colours of wool differed among the Viking-settled areas. Sheep wool in Scotland comes in a full range of colours from black to brown-grey to white. Irish sheep were usually brown and English sheep were white (Henry, 2004). After wool is removed from the animal, it must be carded (combed) to remove impurities (Figure 4.3), then washed and spun.

Yarns were spun on a drop spindle (Figure 4.4) and depending on the size of the spindle, whorl yarns were finer or coarser. For spinning, yarns were twisted clockwise (Z-spinning between the fingers) or anti-clockwise (S-spinning). Z-spinning can produce a firmer thread, but S-spinning is often used for delicate fibres such as silk.

Using different types of wool and spinning yarns in different directions can produce a patterned effect that precludes the laborious work of dying. Textiles in the Viking Age were produced mainly on upright warp-weighted looms (Figure 4.5) that required the wool to be taken over and under the weft. Depending on the sizes of the loom weights, very fine textiles could be produced. A textile is defined as 'fine' when

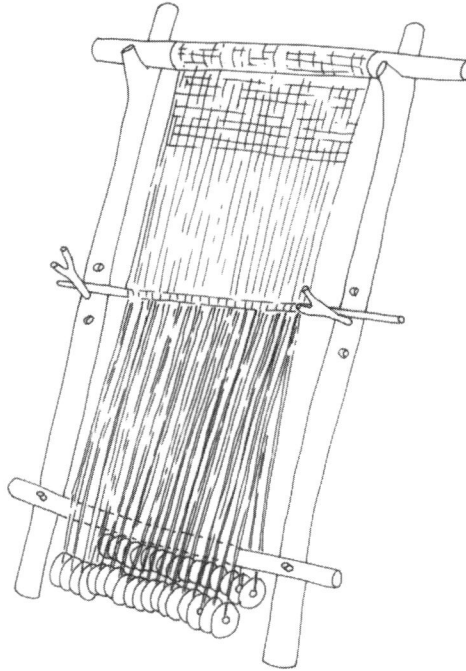

FIGURE 4.5 Warp-weighted loom. Drawing by Annika Jeppsson in Andersson-Strand 2003. (Source: Eva Andersson-Strand, Copenhagen Textile Research Project, http://ctr.hum.ku.dk/. With permission.)

its thread count is high, usually more than 25 threads per centimeter. Various effects achieved by weaving patterns are shown in Figure 4.6.

It is estimated that the wool of 15 to 30 sheep is required to produce about 10 costumes (Andersson-Strand 2002, p. 48). The production of yarn is labour-intensive; to produce enough bedding, clothing and other household goods, it is estimated that a Viking woman would have to work a full 8 months per year on textile production. However, such domestic production pales into insignificance when we consider that about 2000 fleeces were needed to produce the wool for one sail for a Viking ship (Wincott-Heckett 2010).

Not all textiles were produced at home. In the Viking homelands, towns such as Hedeby and Birka established centres of textile production and women may have been organised in guilds. It is also possible that other areas of the Viking world including the Irish Sea region possessed centres of production and that not all textile was homespun.

Flax is an even more labour-intensive material. The plant grows best in damp conditions, which is why Irish linen is still world famous. Flax is demanding of the soil and grows best on light sandy soil free from heavy frosts—a quality that many of the islands around the Irish Sea region, such as the Outer Hebrides, possess. The raw fibre must be retted in standing water to break down the fibres before they can be prepared for use in spinning. Flax must be spun in humid conditions, preferably inside a building; otherwise the yarn will break. It is clear that such a labour-intensive textile can be produced only by settlers who are no longer on the move and have established themselves in new areas.

Comparing textiles from Scandinavian Scotland with those from other areas of the British Isles and Scandinavia gives us an idea of different preferences that tie in with the people who moved to these areas. Philippa Henry (2004) shows most textiles from Scandinavian Scotland are wools and the main weave is tabby made by leading one thread over and under the warp. Few twill weaves were detected, and the main type was a plain four-shed twill (yarn taken over two warps and under two). This weave has parallels in western Norway, but was less popular in Denmark, Sweden, and England. We assume that many settlers to the Irish Sea region had Norwegian backgrounds. These textiles also seem to indicate that the women involved had Norwegian backgrounds. The Veka-style twill that seems to be native to western

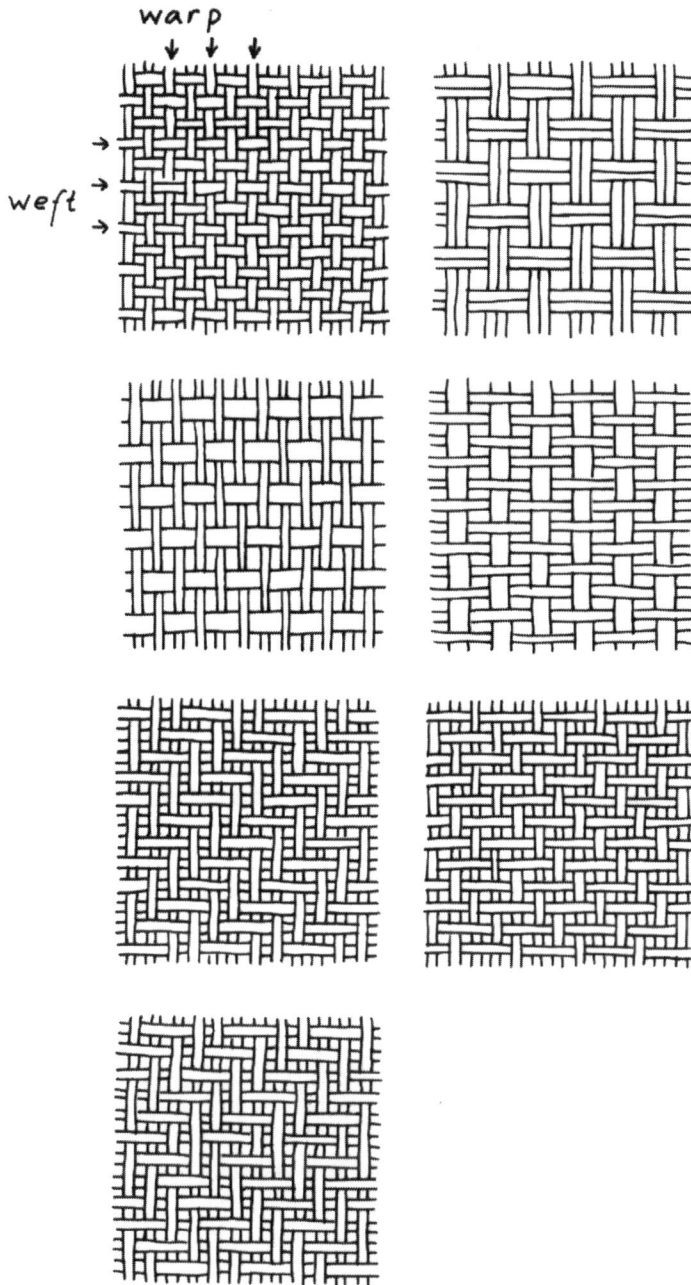

FIGURE 4.6 Textile types. Rows 1 and 2: types of tabby weave. Rows 3 and 4: examples of 2-1 and 2-2 twill weaves. Drawing by Annika Jeppsson. (Source: Eva Andersson-Strand, Copenhagen Textile Research Project, http://ctr.hum. ku.dk/. With permission.)

Norway was found with one male burial in Kildonan on Eigg, Inner Hebrides, but similar examples were found in Dublin and Waterford.

Assumptions of what textiles *should* look like may prevent us from seeing regional differences and adaptations to the new homelands. For example, the well-preserved textiles from Birka in Sweden are usually taken for comparison to Scottish material. The linen of the undershirts worn by Birka women was pleated, but no such pleats have been found on any examples from the Irish Sea region. Similarly, no belt buckles were found at Birka, but we do have an example from Kneep, Isle of Lewis; it was worn over a garment made from Birka-style twill. Belts were worn by English women, so this buckle may have been a variation of dress.

Textiles may underline the connections that Viking settlers made with other areas of the British Isles and Ireland. For example, sprang, a technique of plaiting warps without a weft, is only associated with Scandinavia. It was used when a high degree of elasticity was desired (e.g., for bands that bound trouser legs). An impression on the backside of an oval brooch from a 9th century grave at Clibberswick, Unst, Shetland, has a parallel in 10th century Anglo-Scandinavian Micklegate at York.

Some textiles occur in special contexts in Britain and Ireland. Some fine examples of silk head coverings from Dublin, York, and Lincoln are remarkably similar and may have even been cut from the same bale. This does not indicate that they are forms of ethnic head dresses. Rather, Dublin, Lincoln, and York had close political connections that extended to trade. Imports of such items may have distinguished women of a merchant class who could afford to obtain exotic goods. Textiles may have then, as now, been markers of social status and wearing 'foreign' dress may have been something to which rich women aspired. A wealthy Viking lady may have purchased her clothes from a special market in the same way modern women buy *haute couture* from special retailers.

There is a difference between textiles made in the home and finer fabrics purchased from centres of production. For example, no examples of fine diamond twills (the pairing of the wefts produces a diamond pattern) have been found in the Viking areas of the British Isles other than imprints of metal objects, such as the belt buckle from Kneep that seems to have originated in a larger trading town of Scandinavia.

Other textiles display regional biases. Vaðmál (English Vadmal) was produced in Iceland and across the North Atlantic from the 11th century onward, but there are examples at York and The Biggings, Papa Stour, Shetland. The yarns are Z-spun in one system and S-spun in the other, with coarser wool in the weft. Using different types of wool and spinning methods allowed a fabric to be fulled easily and thus waterproofed. It seems that the origin of this quintessential Icelandic fabric may been the Irish Sea region since early examples were found in Waterford and York (Henry 2004). Z- and S-spun cloth has a strong English and Continental bias and may indicate some form of cross-cultural hybridisation.

Piled fabrics involve the addition of additional tufts of wool during weaving; they were produced in Scandinavia and Ireland. The ground weave in Scandinavia for these fabrics is Z-spun in both warp and weft, but the tufts are S-spun. The fabric has a shaggy appearance. The technique produces a material that is lightweight and semi-waterproof. Irish and Icelandic examples were dyed; all examples from Scotland and the Isle of Man are undyed.

Fabric found at the sword burial of Cronk Moar in the Isle of Man shows an amended technique. The tufts were fastened to the fabric by darning, which gives it a softer appearance. The wool for the cloth was from local sheep. Osteological analysis has shown that the breeds on the Isle of Man varied very little from the Viking Age through today. Such osteological studies of sheep types may therefore tell us about what kinds of raw materials were available to Viking Age weavers.

The distinctiveness of the Isle of Man textile may be notable in connection with other evidence for cross-cultural encounters such as the Hiberno-Norse styles of decoration and the evidence from the Braddan runic inscriptions that suggest Norse-Celtic hybridization. It may even be possible that textiles were 'gendered.' The determination of a male or female burial is usually made on the basis of grave goods, since in many cases skeletal remains are missing or completely decayed. Previous generations of excavators were interested largely in 'treasure' and a lot of organic material was destroyed as a result. However, all instances of piled fabrics found with burials have been gendered male. Most were found in connection with weapons, which, like their Anglo-Saxon counterparts, may have been wrapped in textiles before they were placed into a grave.

Piled fabric was discovered in remains from Anglo-Scandinavian York. It appears, however, that the maker was unfamiliar with the weaving technique, since the tufts were sewn laboriously into the fabric, rather than woven into it. This cloth may represent an imitation of Norse fabric. In contrast to the textile evidence from the Irish Sea region that shows very strong links with the Scandinavian homelands, the textile evidence from the areas of Scandinavian settlement in England such as York have a more

FIGURE 4.7   Reconstructed textile tools: shears, spindle whorl, weaving batten, carding combs, and bone picks. (Source: Eva Andersson-Strand, Copenhagen Textile Research Project, http://ctr.hum.ku.dk/. With permission.)

'English appearance' (Walton 1997, p. 64), possibly indicating that the Anglo-Scandinavian populations adopted each others' textile techniques.

York is an interesting find spot. It has evidence for eastern Scandinavian imports such as honeycomb twills and nålebinding, a form of crocheting developed in Sweden. Nålebinding is virtually unknown outside Scandinavia. Apart from York, the only other example came from Viking Age Dublin in the Irish Sea region. This may suggest that a group of highly skilled textile specialists, perhaps even émigrés of Eastern Scandinavian origin, may have resided in Irish and English Viking Age towns. Still, such tentative connections should not rest on two examples alone because textiles can be transported easily. What these examples show is that trade extended a long way across the Viking world.

Fine textiles may have been raided as well as traded. They may have been used as tax payments to local chieftains and nobles and were thus clearly components of the economy much like metal work and pots. Frankish sources tell us that *opus feminile* (women's work) was paid as a local church tax. Eva Andersson-Strand (2011) recently examined textile production figures from Denmark and Sweden and has shown that textiles were manufactured in towns. Textiles therefore may represent a uniquely female aspect of Viking economy and we should look at the skills and the raw materials that these women used.

Textile tools attributed to high-ranking females such as the whalebone plaques from Sanday in Orkney and the tools from Westness may indicate women were involved in both production and control of the textile industry. We should therefore be careful to associate these objects on the basis of a mythology developed at a much later stage. Some examples of tools are shown in Figure 4.7.

The Westness woman with artefacts from her homelands and those acquired in the new lands demonstrates what material culture can tell us about Viking activities in newly settled places. Her burial as a prominent member of her society underlines how much women shaped the expansion of the Viking world. While literary texts may favour the earls of Orkney and the kings of Dublin, objects also function as texts and tell us about female migration in the Viking Age; we are just now learning how to read them.

Women and Place-Names

Textiles seems to suggest clearly women participated in the Viking expansion to the British Isles and Ireland. Archaeological finds too suggest that Scandinavian women lived and worked around the Irish Sea. However, few text sources explicitly mention Viking women. Two entries in the A version of the *Anglo-Saxon Chronicle* for the year 893 may suggest the presence of Scandinavian women in some of the Viking campaigns.

Of course, these women do not have to be 'Norse' as Judith Jesch (1991; also Chapter 3 of this volume). More conclusive evidence for the presence of Viking women comes from language. Many words in the English language are derived from Old Norse. *Ugly, egg, to die, to kill, skull, sky,* and *window* are all derived from Old Norse. These suggest that communities of speakers continued to speak the language of the homelands in their new home over a considerable period of time.

Place-name evidence indicates that women with Scandinavian names did not only follow their husbands, they also sometimes had enough influence to have places named after them. This is not unusual in England because Anglo-Saxon women could hold lands and often royal ladies had very large estates. The dating of place-names is complex since many of the names only occur in written sources such as *Domesday Book*, which was recorded after the Norman Conquest. Occasional place-names tell us about women's names and indicate how Old Norse may have been influenced by other languages. For example, Bewaldeth in Cumbria combines a *-by* (farmstead) element and the Old English Aldgyth female name (http://kepn.nottingham.ac.uk/map/place/Cumberland/Bewaldeth%20and%20Snittlegarth). Normally a *-by* element follows a personal name, but here the Anglo-Saxon name is preceded by an Old Norse word for 'farmstead,' a trait characteristic of Celtic place names.

While place-names in north-west England often show language contact, a particularly interesting example is the place name of Helperby in neighbouring North Yorkshire. Helperby combines the *Hjalp* element of a female personal place name with the genitive form *-r* and the Old Norse *–by* for farmstead, thus Hjalp's farmstead (http://kepn.nottingham.ac.uk/map/place/Yorkshire%20NR/Helperby). The Old Norse genitive ending indicates that Hjalp lived in a community that spoke Old Norse and spoke it long enough to make sure the combined version became established.

Place-names in north-west England provide really interesting insights about the people who lived there, but they do not necessarily tell us whether these people were Scandinavians. For example, who is the serf in Threlkeld ('thrall's spring')?[*] What place-names do tell us is that Old Norse was used among the speakers of this region.

There are a few other examples of combinations of Old Norse names paired with Old English elements (*tun*, for example, in Aslton (Halfdan's farm) in Cumbria).[†] Carol Hough (2002) observed that this combination was very rare for female names. It may be that Scandinavian women settled in language areas that remained predominantly Norse.

A recent analysis of stray metal finds from female dress fasteners—objects that were lost or discarded—by Jane Kershaw (2009) has shown that despite a corpus of Anglo-Scandinavian styles popular in the areas of Scandinavian settlement in England, there are also objects imported from the homelands. Contrary to the idea that Viking settlers integrated rapidly into English society, dress ornaments continued to emphasise links with Scandinavia.

Conclusion

Unlike the activities of their male counterparts, relatively little attention has been given to women's work in the Viking Age. However, it is clear that women were major contributors to the economy. Whether these women were ethnically Norse or the products of a mixed society remains to be determined. Place-name research in England has demonstrated the presence of female migrants or at least women who were able to speak fluent Old Norse. The same research is needed across other areas of the British Isles. It is clear that Viking women shaped the new lands they settled economically, socially, and linguistically, and they deserve places right next to their men.

ACKNOWLEDGMENTS

This essay is based on a paper written for a public engagement event and retains this character. I would like to express my thanks to the Chester Viking Conference organisers and the audience. I am sincerely grateful to Eva Andersson-Strand (Danish National Research Foundation's Centre for Textile Research

---

[*] http://kepn.nottingham.ac.uk/map/place/Cumberland/Threlkeld
[†] http://kepn.nottingham.ac.uk/map/place/Cumberland/Alston%20with%20Garrigill

at University of Copenhagen), who provided the images for this chapter. Research for this paper began a long time ago when I worked for the Manchester Medieval Textile Project and I am very grateful to Betty Coatsworth for letting me use the database for the Isle of Man textiles.

Some very good scholars are looking at Viking Age textiles now. I would like to urge readers to consider the work of my colleagues cited in the references. Readers may also be interested in the websites of the Anglo-Saxon Laboratory, which is the leading textile research centre in the United Kingdom [http://www.aslab.co.uk/] and the Danish National Research Foundation's Centre for Textile Research [http://ctr.hum.ku.dk/] for further reading. For place-name research, the publications of the English Place Name Society are important and the *Key to English Place-Names* [http://kepn.nottingham.ac.uk/] of the Institute for Name Studies is a very valuable tool.

REFERENCES

Andersson, E. (2003). Textile Production in Scandinavia during the Viking Age. In Bender-Jorgensen, L. et al., Eds., *Textilien aus Archaeologie und Geschichte: Festschrift für Klaus Tidow.* Neumünster, K. Wachholz, pp. 46–62.

Andersson, E. (1999). *The Common Thread: Textile Production during the Late Iron Age–Viking Age.* Institute of Archaeology Report Series, 67, Lund, Sweden.

Andersson-Strand, E.B. and Nosch, M.L.B. (2013). *Tools, Textiles, and Contexts.* Ancient Textiles Series, Vol. 13. Oxford, Oxbow Press.

Andersson-Strand, E.B. (2011). Tools and Textiles: Production and Organisation in Birka and Hedeby. In *Viking Settlements and Viking Society: Proceedings of 16th Viking Congress,* Sigmundsson, S. and Holt, A., Eds. Reykjavik, Hið Íslzka Fornleifafélag and University of Iceland Press, pp. 1–17.

Andersson-Strand, E.B., Frei, K.M., Gleba, M. et al. (2010). Old Textiles, New Possibilities. *European Journal of Archaeology* 13, 149–173.

Geijer, A. (1938). *Die Textilfunde aus den Gräbern.* Uppsala, Almqvist & Wiksell.

Graham-Campbell, J.A. (1975). Two Scandinavian Brooch Fragments of Viking Age Date from the Outer Hebrides. *Proceedings of Society of Antiquaries of Scotland* 106, 212–214.

Hägg, I. (1991). *Die Textilfunde aus der Siedlung und aus den Gräbern von Haithabu.* Neumünster, K. Wachholz.

Henry, P. (2004). Changing Weaving Styles and Fabric Types: Scandinavian Influence. In Hines, J., Lane, M., and Redknap, M., Eds., *Land, Sea, and Home.* Leeds, Maney, pp. 443–456.

Hough, C. (2002). Women in English Place-Names. In Hough, C. and Lowe, K.A., Eds., *Lastworda Betst: Essays in Memory of Christine E. Fell with her Unpublished Writings.* Donnington, Shaun Tyas, pp. 41–106.

Ibn Fadlān (2012). *Ibn Fadlān and the Land of Darkness: Arab Travellers in the Far North.* Translated with introduction by Lunde, P. and Stone, C. London, Penguin.

Jesch, J. (1991). *Women in the Viking Age.* Woodbridge, Boydell Press.

Kershaw, J. (2009). Culture and Gender in the Danelaw: Scandinavian and Anglo-Scandinavian Brooches. *Viking and Medieval Scandinavia* 5, 295–325.

Lee, C. (2014). Crossing the Irish Sea: Female Migration and the Viking Age. *Journal of the Royal Society of Antiquaries of Ireland* (in press).

Owen-Crocker, G. (2004). *Dress in Anglo Saxon England: Revised and Enlarged Edition.* Woodbridge, Boydell Press.

Sutherland, P. (2009). The Question of Contact between Dorset Paleo-Eskimos and Early Euroepans in the Eastern Arctic. In Maschner, H., Mason, O., and McGhee, R., Eds., *The Northern World,* pp. 279–299.

Walton, P. (1997). Textile Production at 16-22 Coppergate. *The Archaeology of York: The Small Finds.* York, York Research Trust.

Walton, R.P. (2007). *Cloth and Clothing in Early Anglo-Saxon England, AD 450–700.* York: Council for British Archaeology.

Williams, H. (2006). *Death and Memory in Early Medieval Britain.* Cambridge, Cambridge University Press.

Wincott, H.E. (2010). Textiles that Work for a Living: A Late 11th Century cCoth from Cork, Ireland. In Sheehan, J. and O'Corráin, D., Eds. *The Viking Age: Ireland and the West.* Dublin, Four Courts, pp. 555–564.

5

Taking Sides: North-West England Vikings at the Battle of Tettenhall, AD 910

John Quanrud

CONTENTS

ABSTRACT The events recorded in the Anglo-Saxon Chronicle for the years 909 and 910 resulted in a great English victory over a Scandinavian enemy at Tettenhall, Staffordshire that has long been regarded as a major turning point in the history of Anglo-Saxon England. Since the late 11th and 12th centuries, it has largely been assumed that these annals refer to the Northumbrian Danes based at York. Their defeat in 910 is seen as marking the beginning of a new phase of successful English military campaigning leading to the 'reconquest' of the Danelaw—a progressive series of conflicts and capitulations in the Scandinavian-ruled territories of eastern England culminating in the so-called 'submission' of the northern kings and rulers to Edward the Elder at Bakewell in 920.

It will be suggested that this traditional interpretation of the battle of Tettenhall, despite its near-universal acceptance, may be flawed. The annals in the Chronicle will be assessed, the origins of the tradition considered, and (based on the significant and growing evidence related to Irish Sea Viking presence in north-west England then), an alternative will be presented that, if deemed viable, may have considerable implications for interpreting the subsequent history of insular politics in the early 10th century.

Introduction

Within the narrative framework of the Anglo-Saxon Chronicle (ASC), the battle of Tettenhall precedes an unexplained but undeniably pivotal shift in English and Scandinavian relations in Britain. Sir Frank Stenton argued that the defeat suffered by the Scandinavians in 910 created a major imbalance of power and sparked a decade of unrest affecting all the major kingdoms of Britain. In his view, '[t]he Danish armies in Northumbria never recovered from the disaster of 910'—a disaster that 'opened the way to the great expansion of the West Saxon kingdom in the following years. The gradual reduction of East Anglia and the Danish midlands by the king of Wessex could never have occurred if the Danish colonists of that country had been supported by a strong Northumbrian kingdom' (Stenton 1971, pp. 323–324).

F. T. Wainwright fully agreed with this assessment, describing the battle as 'one of the most significant events of [Edward's] reign.' He concluded, '[i]t would be difficult to exaggerate the calamity suffered by the Danish power at Tettenhall' (Wainwright 1975, p. 88). It is this interpretation, this received tradition that a weakened Northumbrian kingdom paved the way for English aggression against Danish-ruled lands further south that served as a cornerstone of understanding, not only for numerous historical works dealing with 10th century Britain, but across all the disciplines from archaeology and numismatics to stone sculpture analysis, place-name studies, and more. But is it correct?

Anglo-Saxon Chronicle and the Battle of Tettenhall

What, then, is known about the battle of Tettenhall? According to ASC annals 909 and 910,* King Edward sent a combined force of West Saxons and Mercians to raid amongst a 'northern army,' killing many. The Scandinavians retaliated the next year with a foray deep into Mercian territory (Figure 5.1). On their return home, they were overwhelmed by the sudden and unexpected arrival of Edward's men. A battle ensued at Tettenhall (near modern-day Wolverhampton) where the English won a significant victory over the Vikings with immense slaughter. The story in MS C of the Chronicle† reads (O'Brien-O'Keeffe 2001, p. 73: 910 for 909, and 911 for 910):

> 909 … 7 þy ilcan gere sende Eadweard cing fyrde ægþer ge of Wessexum ge of Myrcum, 7 heo gehergode swiðe micel on þam norðhere ægþer ge on mannum ge on gehwilcum yrfe, 7 mænige menn ofslogon þara deniscra 7 þær wæron.v. wucan inne.

> 910 Her bræc se here þone frið on Norðhymbrum 7 forsawan ælc riht þe Eadweard cing 7 his witan him budan 7 hergodon ofer Myrcna land. 7 se cing hæfde gegadorod sum hund scypa 7 wæs ða on Cent, 7 þa scipu foron be suþan east andlang sæ togenes him. Þa wende se here þæt his fultomes se mæsta dæl wære on þam scypum 7 þæt hi mihton unbefohtone faran þær ðær hi woldon. Þa geahsode se cing þæt þæt hi on hergeað foron, þa sende his fyrde ægþær ge of Wessexum ge of Myrcum, 7 hie offoron þone here hindan, þa he hamweard wæs, 7 him wið gefuhton 7 ðone here geflymdon 7 his fela ðusenda ofslogon. 7 þær wæs Eowils cing ofslegen, 7 Healfden cing, 7 Ohter eorl, 7 Scurfa eorl, 7 Oðulf hold, 7 Benesing hold, 7 Anlaf se swearta, 7 Þurferð hold, 7 Osferð hlytte, 7 Guðferð hold, 7 Agmund hold, 7 Guðferð.

In one of the most commonly used translations of the Chronicle, Dorothy Whitelock, David Douglas, and Susie Tucker (1961, pp. 61–62) rendered these annals as follows:

---

* ASC annals are cited according to the years assigned in the translation by Whitelock et al.

† Several versions of ASC have survived, and each has been designated a single letter to distinguish between them—MS A, MS B, MS C, and so on. When referencing multiple ASC manuscript versions, the abbreviation 'MSS' is used, followed by the relevant letters, for example, 'MSS BCDE', or simply 'BCDE'.

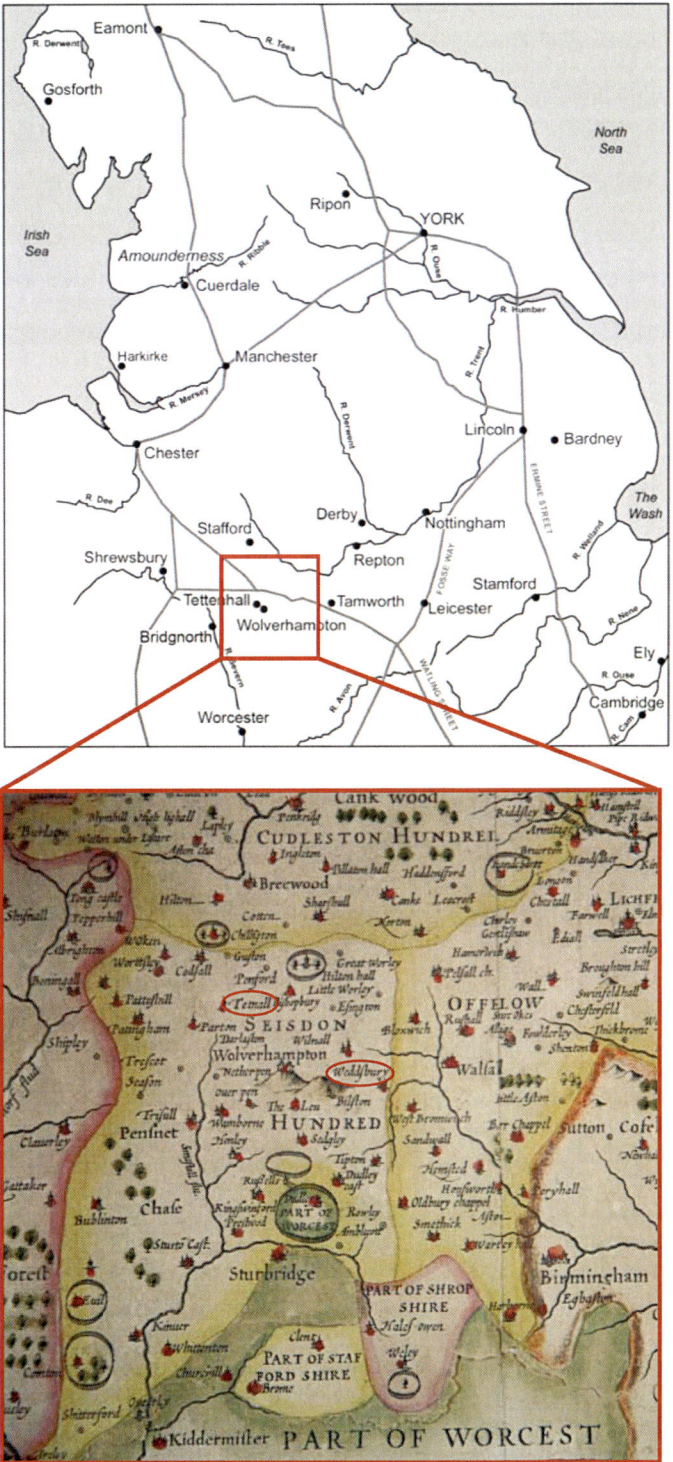

FIGURE 5.1 Map showing Tettenhall and Wolverhampton. Expanded section is part of a 1611 map of Staffordshire by John Speed showing Tettenhall as Tetnall, and Wednesbury (reproduction by permission of the William Salt Library, Stafford).

909 ... And that same year King Edward sent an army both from the West Saxons and from the Mercians, and it ravaged very severely the territory of the northern army, both men and all kinds of cattle, and they killed many men of those Danes, and were five weeks there.

910 In this year the army in Northumbria broke the peace, and scorned every privilege that King Edward and his councillors offered them, and ravaged over Mercia. And the king had collected about 100 ships, and was then in Kent, and the ships were on their way south-east by sea towards him. Then the Danish army thought that the greater part of his forces was on the ships, and that they could go unopposed wherever they wished. Then the king learnt that they had gone on a raid. He then sent his army both from the West Saxons and Mercians, and they overtook the Danish army when it was on its way home and fought against it and put the army to flight and killed many thousands of its men. And there were killed King Eowils and King Healfdene and Earl Ohter and Earl Scurfa, and Othulf the *hold*, and Benesing the *hold*, and Olaf the Black and Thurferth the *hold*, and Osfrith Hlytta, and Guthfrith the *hold*, and Agmund the *hold* and Guthfrith.

At first glance, the meaning of these passages may seem obvious, but a closer look reveals considerable ambiguity. The Old English (OE) terms *norðhere* and *Denisc* translated here as 'northern army' and 'Danes' are not, in fact, specific in their identification of the Scandinavian force or its location. Indeed, the lack of clarity is such that certain assumptions must be brought to bear from the wider historical context for these annals to be fully understood. And herein lies the challenge for, as Alistair Campbell (1942, p. 85) once observed, '[a]fter the death of the ætheling Æthelwald at the battle of the Holme and the conclusion of the peace of [Tiddingford] with the Norsemen of both East Anglia and Northumbria, the affairs of Northumbria are shrouded for several years in complete darkness.'*

Central to the problem of interpreting these texts are matters related to this 'northern army.' Where was it located? Who were its leaders? What was its relationship to the other political powers of early 10th century Britain?

Starting Point

Since the days of the Anglo-Norman historians it has been commonly assumed that the *norðhere* that Edward's army 'ravaged very severely' in 909 was the Danish army of Northumbria with its base of power centred on York. For the Anglo-Normans who were largely (and often solely) dependent upon the texts of the Chronicle for their knowledge of the Anglo-Saxon period, this was the only interpretation available to them. To this point in ASC, specific reference to Scandinavian settlement in England is limited to regions north and east of Watling Street, the border established by Alfred and Guthrum in the treaty they reached between 878 and 886.

It may now, however, be possible to consider an alternative interpretation. Since the mid-19th century and through the ongoing study of place-names, stone sculpture, silver and/or coin hoards, burials, grave goods, and other archaeological finds in north-west England, acceptance has grown among scholars that by the first decade of 10th century another Scandinavian immigration (ignored by ASC's chroniclers and quite separate from the earlier one in eastern England), was underway in north-western regions (Wainwright 1975, pp. 182–184; Edwards 1998; Andersen 2006, pp. 9–58; Graham-Campbell and Philpott 2009).

One assumption fundamental to the arguments that follow is that the early years of Edward the Elder's reign saw the start of wide-scale Viking settlement in north-west England. What evidence there is suggests that this immigration was, in part at least, related to the expulsion from Dublin and other regions in Ireland *c*. 902 of large numbers of Scandinavians under the leadership of Hiberno-Norse kings (Wainwright 1975, p. 219).

Colmán Etchingham has argued that the Vikings whose base of power was centred on Dublin from the mid-9th century and who are at times referred by Irish annalists as OI *Finngaill* (white foreigners) were

---

* The battle of the Holme took place *c*. 902; the peace at Tiddingford in 905 or 906.

'primarily Norwegian' and distinguished by the Irish from the *Dubgaill* ('black foreigners'). This second term, according to Etchingham, 'most commonly denotes Vikings active in Britain, who were primarily Danes and who intervened in Ireland in 851-2 and 875-7' (Etchingham, 2014, p. 37). Etchingham here counters a theory put forward by David Dumville (2005, pp. 78–93) and more recently developed by Clare Downham (2007, pp. xvi–xx, 14–15 and 22–23; 2009, pp. 150–152) that *Dubgaill* refers to a family dynasty that established itself in Dublin in the early 850s and expanded its rule to York in the 860s, creating a sort of Scandinavian axis between these two centres of power.[*]

ASC Annal 909

What does ASC have to say about the enemy that Edward the Elder sent his men to harry in 909? The term *se here* (the army or the raiding army) often appears in ASC on its own or together with words such as *micel* (great) *scip* (ship), and *hæðen* (pagan) to describe the various Scandinavian forces encountered by the Anglo-Saxons in the 9th and 10th centuries.[†] Annal 909, however, contains the only occurrence of the compound *norðhere* in the entire Chronicle. This fact limits the usefulness of comparative linguistic analysis in this instance.

As to which army is thus indicated, the *norð* in *norðhere* either depicts its location, as in 'the army to the north' or 'northern army' or refers to its members, as in an 'army of *norð*-men' in much the same way that *sciphere* (army of ships) denotes a sea-borne force and *hæðen here* indicates an army of pagan warriors. It is clear from this annal's context that the *norðhere* was based in Britain, and from annal 910 that it was to the north of Mercia.

Nothing in either annal, however, helps narrow down the geographical boundaries of the *þær … inne* where the English carried out their five-week raid. No mention is made of York or the Danes of York.[‡] It is therefore possible that *norðhere* was used in annal 909 to introduce a Viking army that was new on the scene (such as, perhaps, the Irish Sea Vikings in north-west England) and therefore needed to be distinguished from other Scandinavian armies.[§]

OE Denisc, Dene and Norðmenn in ASC annal 909 also reports that the members of the *norðhere* who fell as a result of the English raids were *þara deniscra*, which Whitelock et al. designates as 'those Danes.' To a modern reader the terms *Dane* and *Danish* imply an ethnic identity that was almost certainly not intended by ASC's chroniclers as distinguishing Danes from Norwegians or others with ties to Scandinavia, whether past or present.

It has been suggested that *Denisc* in this adjectival form may have been used in ASC in the sense of 'Scandinavians in general' (Downham, 2007, p. xviii). This notion (at least as far as the Common Stock is concerned, the annals of which continue to the last decade of the 9th century) may find some support in ASC 789 in which MSS BCDE agree that *iii. scipu Norðmanna* (three ships of Northmen) sailed, according to the *Annals of St Neots* (Dumville and Lapidge 1985, p. 39) – into Portland in Dorset, where they murdered the king's reeve.

MSS DE add that these intruders were *of Hæreðalande*, described by Peter Sawyer (1962, p. 2) as 'Hörthaland, a district of western Norway'. Annal 789 continues with the statement that these ships from the western coast of Norway were *ærestan scipu Deniscra monna þe Angelcynnes lond gesohton* (the

---

[*] Etchingham's paper will be considered in more detail below.

[†] See, for example, annals for 860, 865, 866, 875, 877, 885, and 893.

[‡] York is named in ASC's Common Stock, for example, at 867 (ADE *Eoforwicceastre on Norþhymbre*; BC *Eoforwicceastre on Norðanhymbre*); 869 (ABCD *Eoforwicceastre*; E *Eoferwicceastre*); and 919 (D *Eoforwic*; E *Eoferwic*).

[§] Colmán Etchingham writes, '[g]iven that *Norðmanni* is used in the Irish annals of the pre-851 element otherwise identifiable as *Finngaill*, and which, as I hope I have shown, were those who controlled Dublin after 853, could it be that *Norðmanni*, although evidently interchangeable with *Dani* in Frankish chronicling convention, carried the specific connotation of 'Norwegian' Vikings for Insular chroniclers, both Irish *and* English (who used vernacular *Norðmanni* in 789) who, to a greater extent than the Franks, were confronted with *both* 'Danes' *and* 'Norwegians'? Could such a specific 'Norwegian' connotation be echoed in *norðhere*? Could *Norðmanni* of ASC 789 point to the inspiration for the Hiberno-Latin annalistic usage 837–948, the first example of which is, of course, not, in fact, Hiberno-Latin, but Irish?' (personal communication).

first ships of the Danish men which sought out the land of the English race), This final phrase also was included in MS A (Bately 1986, p. 29: 787 for 789; Whitelock et al. 1961, p. 35).

In this instance, *Denisc* clearly does not refer to men who came to England from Danish lands; they came from what is Norway today. That *Norðmenn* are here described as *Denisc* does not mean, however, that these terms were interchangeable, for *Denisc* is the only adjectival form used in ASC. Also noteworthy is the fact that the use of *Denisc* in the Common Stock annals is oddly uniform, for, apart from this lone example of *Norðmenn* in annal 789 in BCDE, the nouns *Dene* and *Norðmenn* never occur in the annals of the Chronicle to the end of the 9th century.[*]

Both terms appear in other OE sources, for example in the Old English *Orosius* which dates to the reign of Alfred the Great.[†] This suggests that the form *Denisc* was used to the exclusion of both *Dene* and *Norðmenn* and may therefore represent an editorial decision taken by ASC's authors when compiling that work rather than reflecting standard usage over decades, or indicating, as Downham (2009, p. 143) suggested, that Anglo-Saxons in this period may not have attempted 'to distinguish between Viking-groups of "Danish" and "Norwegian" ethnicity.'

In what may be a significant observation in this regard, Cyril Hart noted that in the appendix that forms part of the *Chronicle of John of Worcester* (or the *Chronicon ex chronicis*), the pagan force active in England in 867 is reported as having included 'Danis, Norreganis, Suanis, Goutis et quarundum aliarum natione populis,' an identification which 'is not based on any known source' (Hart 1983, p. 277).

The first occurrence of *Norðmann* after 789 is in annal 920 (the 915 to 920 section is found only in MS A), and so belongs to one of the continuations added to ASC after the original Common Stock was completed. The *Denisc* description in annal 909 for the men of the *norðhere* killed by Edward's army cannot, therefore, be taken to mean Danish in the modern sense of the word, or, more specifically, as the 'Danes of York.'

Edward's Motive for Sending His Army in 909

Given that annal 909 fails to identify the origins or contemporary location of the Scandinavian 'northern army' that caused the Anglo-Saxons such concern, it remains to consider why Edward sent out his *fyrd* in the first place. Unfortunately, annal 909 offers no hint as to a possible motive. In the context of the Chronicle narrative, the raid occurs suddenly and without warning. Edward's previous dealings with Scandinavians had been in 905 or 906 when, according to MSS ABCD (Bately 1986, p. 63: 905; Whitelock et al. 1961, p. 209): '…mon fæstnode þone frið æt Yttingaforda, swa Eadweard cyng gerædde, ægðer wið Eastengle ge wið Norðhymbre.' (…the peace was established at Tiddingford, just as King Edward decreed, both with the East Angles and the Northumbrians.)

MS E states, on the other hand, that the Tiddingford treaty was established not 'as King Edward decreed', but *for neode* 'from necessity' (Irvine 2004, p. 54: 905). Whatever the purpose behind Edward's *frið* with the kings of East Anglia and Northumbria *c.* 905, a dramatic change must have occurred in the three or four years that followed, either in Edward's strategic planning or in the wider Scandinavian political landscape. All that is known for certain from annal 909 is that Edward sent an army of West Saxons and Mercians to harry a 'northern army' in the territory in which it was settled, that the operation lasted five weeks, and that the English destroyed much property and killed many Scandinavians.

Any further interpretation of this annal, such as equating the *norðhere* with the Danes at York, is conjecture, for the text makes no mention of York and lacks sufficient detail to determine more precisely who, where, or why the English were fighting in 909. Indeed, the history of Britain and its peoples in this period presents many challenges to scholars, as Wainwright (1975, p. 163) observed:

> A heavy mist hangs over the north. We do not know what happened to the shattered fragments of
> the Anglian kingdom after the battle at York on 21 March 867; we know little about the Danish

---

[*] *Denisc* occurs over 20 times in MS A between 789 and 896.

[†] Irmeli Valtonen (2008, p. 344) wrote of the *Orosius* translation that it 'is the earliest OE text where the name *Dene* occurs, unless *Widsith* and *Beowulf* or their sources are dated earlier.'

kingdom later established, and we know even less about the subsequent relations of the Angles, the Danes and the other peoples of the north. From about 900 onwards Norsemen from Ireland poured into north-western England, and the expedition of Ragnald* may well mark the culmination of this movement. Another element was thus added to the racial complex, and though we may speculate on possible repercussions we can be sure only that the arrival of the Norsemen disturbed whatever uneasy political balance then existed.

It is a curious fact that the uncertainty Wainwright expresses here is rarely reflected in the historical works and references dealing with the events of 909 and 910. Nearly every historian who has written about the raid of 909 or the subsequent battle in 910 has invariably repeated, in one form or another, that it was to York or territories under the control of that kingdom that Edward sent a joint force for five weeks, and the Danes and kings of York died at Tettenhall (Stenton 1971, pp. 323–324; Whitelock 1979, p. 33; Sawyer 1998, pp. 116–118).† While this may be a correct interpretation of the evidence, it should be recognized as such—an interpretation based on a certain set of assumptions.

ASC Annal 910

Returning to the texts of ASC, annal 910 (MSS BCD) begins by reporting that *se here* (presumably the one the English attacked in 909) violated *þone* frið *on Norðhymbrum* (the peace in Northumbria)—a treaty for which no further information is given—and raided deep into Mercia with disastrous consequences. There are a number of variations between the different versions of this annal. For example, MS A (Bately 1986, p. 63: 911 for 910; Whitelock et al. 1961, p. 61) begins: 'Her bræc se here on Norðhymbrum þone frið, 7 forsawon ælc frið þe Eadweard cyng 7 his witan him budon…' (In this year the army in Northumbria broke the peace, and scorned every privilege that King Edward and his councillors offered them….)

The first variation involves a discrepancy as to which noun the phrase *on Norðhymbrum* (in Northumbria) is intended to qualify. According to MS A it was *se here on Norðhymbrum* (the army in Northumbria) that broke the peace, while in BCD it was *þone* frið *on Norðhymbrum* (the peace in Northumbria) was broken by *se here*. If the scribe of MS A correctly copied the original annal and it read *se here on Norðhymbrum*, then, as this same phrase is used to describe the Danish army of York in the annal for 900, it might be argued that here too the text refers to that same army.‡

If, however, the reading of MSS BCD is correct and the peace in Northumbria had been violated by *se here* (the *norðhere* of annal 909), then it may refer to a Scandinavian army based elsewhere in the north. In actual fact, it probably makes little difference either way. If the Scandinavian army attacked by the English in 909 that retaliated in 910 was made up of the newly established Irish Sea Vikings in western Northumbria, such a force could equally have been described by an English chronicler in the early 10th century as either the *norðhere* or *se here on Norðhymbrum*. To an Anglo-Saxon author at that time, *Norðanhymbre* could be used for any or all of the territories that once belonged to that former kingdom, whether to the east or west of the Pennines.§

Although the Anglo-Saxon kingdom of Northumbria came to an end with the fall of York in 867, use of the term *Norðanhymbre* to describe its former territories that had stretched from the Irish Sea to the North Sea continued long after its demise. For example, when Edward in 919 sent his

---

* It is reported in MSS DE that Ragnald captured York in 919 (Irvine 2004, p. 54: 923 for 919).
† Nick Higham (1992, p. 24) suggested that the northern army may have been based in Lancashire (albeit as a force subject to the rulers of York) and that the 909 raids may therefore have taken place in territories west of the Pennines.
‡ In a bid to bolster his claim to the throne, Edward's cousin and rival, Æthelwold, in 900 *gesohte þone here on Norðhymbrum* (went to the Danish army in Northumbria) (O'Brien 2001: 901 for 900; Whitelock et al. 1961, p. 59).
§ ASC's annals for 893, 894, 896, and 905–6 use *Norðanhymbre* together with *Eastengle* to describe the two Scandinavian-ruled kingdoms north and east of Watling Street. The question of who controlled the Northumbrian territories to the west of the Pennines after 867 will be considered below.

men 'to repair and man' Manchester, the chronicler refers to it as *Mameceaster on Norþhymbrum* (Manchester in Northumbria) (Bately 1986, p. 69: 922 for 919; Whitelock et al. 1961, p. 67).[*]

It would seem that the phrase *se here on Norðhymbrum* in MS A together with the *Denisc* term describing the men killed by the English in 909 served to strengthen the impression among scholars that the Scandinavian army of 910 is synonymous with the Danes of York. On the other hand it might be asked, if this annal is introducing a new army into the mix, then why is it not made more obvious? The fact is, however, that nothing specific was ever reported in the Chronicle concerning the influx of Scandinavians into north-west England in the early 10th century despite the considerable evidence suggesting that such a migration did indeed take place.

The second variation in annal 910 is also between MS A and the other versions and it is found in the continuation of the first sentence. MS A reports that the Scandinavian army *forsawan ælc frið* (scorned every peace or scorned every privilege). According to BCD, *ælc riht* (every right) had been scorned. Cyril Hart suggested that the compiler of MS A may have amended the text of the archetypal Chronicle source to enhance King Edward's reputation beyond what appeared in the original.

Hart (1982, p. 254) cites examples such as the use of *frið* for *riht*, among others, and argues that in this instance the copyist 'preferred to say that King Edward offered "privileges" to the Northumbrians rather than "rights."' According to the *Anglo-Saxon Dictionary*, *riht* can be defined as 'what properly belongs to a person, what may justly be claimed, a 'right,' or 'due' (Bosworth and Toller 1898, p. 796). A *frið* (*ibid.*, p. 338) was something which was in the king's power to agree or bestow. If Hart is correct, and the scribe responsible for MS A intentionally modified the text as he copied from his source, it is possible that for whatever reason including scribal error, *se here on Norðhymbrum* in MS A replaced the original *þone frið on Norðhymbrum* as found in the other vernacular versions.

Broken Peace of 910

It remains to consider the nature of the *frið* that the raiding army violated in 910. The term may refer to various alternatives, but again the Chronicle is silent. It could refer to the Tiddingford treaty of 905–6 which Edward agreed 'both with East Anglians and with Northumbrians.' More compelling perhaps is the possibility that it has to do with conditions imposed by the English on the Scandinavians as a result of the 909 raid. ASC, however, does not report on the establishment of such an agreement.

Simon Keynes (1986, p. 199) writes that treaties with Viking armies often involved 'the exchange of hostages, the reciprocal swearing of oaths and agreement of terms, in different combinations according to the particular circumstances, sometimes with further refinements such as baptism of the Viking leaders.'

It is certainly feasible that as a result of the 909 raids a *frið* was drawn in favour of the English, addressing the issues for which Edward dispatched his men against the *norðhere*. Æthelweard, who compiled his *Chronicon* in the second half of the 10th century, also mentions a treaty in 910 but adds an intriguing and in this regard possibly significant statement concerning Æthelred, the lord of the Mercians. It reads (Campbell 1962, p. 52): 'Annum post unum barbari pactum rumpunt Eaduuardum regem aduersus, nec non contra Ætheredum, qui tum regebat Northhymbrias partes, Myrciasque.' (After a year the barbarians broke the peace with King Eadweard, and with Æthelred, who then ruled the Northumbrian and Mercian areas.)

It is noteworthy that Æthelweard names Æthelred here, for ASC 910 reports only Edward, together with his *witan*, as having negotiated with the Vikings, effectively obscuring any role the Mercian leader may have played in the proceedings. Why does Æthelweard name Æthelred at this point in his *Chronicon*? Did he merely assume Æthelred's participation or was he using a now lost source? Æthelweard could not have extrapolated from the Chronicle account that Æthelred was involved in negotiating the treaty the Scandinavians violated in 910 or that he 'then ruled the Northumbrian and Mercian areas.'

---

[*] Annal 920 (MS A) reports that from among the Northumbrians, *ge Englisce ge Denisce ge Norþmen* (Bately, 1986, p. 69: 924 for 920) acknowledged Edward as father and lord at the Bakewell meeting. It has been argued that *Norþmen* here may refer to those who, with Ragnald, recently captured York (Smith 1928, p. 199; Wainwright 1975, pp. 341–343). Clare Downham (2009, pp. 143–146), however, has argued against this interpretation.

That Æthelweard had access to sources other than ASC is apparent in other parts of his *Chronicon* (Campbell 1962, p. 31; Keynes and Lapidge 1983, p. 189; Bately 1986, pp. lxxx–lxxxi), and appears to be the case in this instance. It is unfortunate, however, that Æthelweard's statement is not specific; the precise Northumbrian location to which the phrase concerning Æthelred, *qui tum regebat Northhymbrias partes, Myrciasque* (who then ruled the Northumbrian and Mercian areas) refers to is uncertain. If it is true that Æthelred ruled the 'Northumbrian and Mercian areas,' it is difficult to see how the Mercian leader could have exercised authority over the Danish kingdom of York in any way comparable to his position in Mercia.

Although Æthelweard's Latin has attracted much criticism (Whitbread 1959, p. 579), his use of sources has been shown on the whole to be trustworthy. It would appear, therefore, that (1) Æthelweard exaggerated Æthelred's influence in this regard, (2) the 909 raid resulted in some form of capitulation at York, or (3) the 'Northumbrian areas' phrase refers to Æthelred's rule over the western territories of that former kingdom.

While the first option is possible, the second is unsubstantiated and wholly untenable, not least because, had Edward's army imposed a *frið* on York in 909, it is certain from the way his other activities were reported by Chronicle author(s) to 920 that such a success would have been spelled out in the clearest of terms (Pelteret, 2009). Some support for the third option may be found in Stenton's suggestion (1970, p. 216) that '[t]here is no reason to think that the English occupation of the north-west was affected by the decline of Northumbrian power in the 8th century, and the little evidence which exists suggests that the country west of the Pennine hills was still subject to English rulers when King Alfred died in 899.'

This assessment, together with Æthelweard's claim regarding Æthelred's rule in Northumbrian parts, may suggest that western Northumbria came under some form of Mercian control in the years after the collapse of Anglo-Saxon Northumbria in 867, this having been the Northumbrian 'area' over which Æthelred exercised some authority. If this is correct, then it may be that the *frið on Norþhymbrum* violated by the Scandinavian army in 910 involved the north-west and not the north-east of England and subsequently Edward sent his army *not* to York in 909, but instead against the territories of the newly established *norðhere* which, as well as parts of Cheshire, may have controlled bases in Lancashire and Cumbria.

Such an interpretation would fit with the report in the *Fragmentary Annals of Ireland* –about ongoing unrest in north-west England which states that not long after Ingimund's attack on Chester 'there was fighting again' (Griffiths, Chapter 2).[*] Similar unrest probably led Æthelflæd to fortify Chester in 907 (Taylor 1983, p. 49; O'Brien-O'Keefe 2001, p. 75; Whitelock et al. 1961, p. 61; Wainwright 1975, pp. 84–86), and may even have caused the removal of St. Oswald's remains from Bardney (Lincolnshire) to Gloucester in 909.[†]

Other evidence beyond that found in these texts may help determine whether the conflicts of 909 and 910 played out in western Northumbria. However, it is first necessary to consider the final variation in ABCD's annal 910, one that may also suggest that Irish Sea Vikings in north-west England were behind these events.

Scandinavian Kings Killed at Tettenhall

Apart from the usual minor differences in dialect, the rest of the Chronicle annal for 910 is the same in ABCD down to the concluding list of names of the Scandinavian leaders who lost their lives at Tettenhall. Considerable discrepancies exist among the versions. In MS A, surprising perhaps, given this text's apparent propensity to enhance Edward's reputation where it can, only one name is recorded from among the 'thousands' of fallen Scandinavians: *Ecwils cyng* (Bately 1986, p. 64).[‡] The compiler of

---

[*] Wainwright (1975, p. 313) wrote that 'the primary function of the fortress at Chester was to overawe the Irish-Norwegian settlers in Wirral.' Griffiths (2001) argued that Chester and its western neighbouring burh of 921, Cledemutha, were there to dominate the Welsh borderlands and control the Dee and its connection to the Irish Sea, but this interpretation assumes a far less substantial Irish Sea Viking presence in north-west England than is suggested in this paper.

[†] Peter Sawyer (1998, p. 117) suggested that St. Oswald's remains were taken from Bardney by force during the English raids of 909, but the report in the *Mercian Register* for 909 says nothing of violence and gives no reason for this action.

[‡] OE *Ecwils* also appears as *Eywysl, Eowils,* and *Auisle* in the different sources (Dumville, 2004, p. 84, note 33).

MS D, on the other hand, adds five more, *Healden cyng, Ohter eorl, Scurfa eorl, Aþulf hold*, and *Agmund hold* (Cubbin 1986, p. 38). MSS BC include all these plus another six, *Benesing hold, Anlaf se swearta, Þurferð hold, Osferð Hlytte, Guðferð hold,* and *Guðferð*, a total of twelve in all (Taylor 1983, p. 47; O'Brien-O'Keeffe 2001, p. 73). Æthelweard, in his *Chronicon*, mentions three kings in his summary of the battle of Tettenhall which reads (Campbell 1962, p. 53):

> Ibidemque ruunt reges tres eorum turbine in eodem (uel certamine dicere fas est) scilicet Healfdene, Eyuuysl quoque, nec non Iguuar relicta tyrannide tum ad aulam properauit inferni, maioresque natu eorum, duces ac nobiles simul. (There fell three of their kings in that same 'storm' (or 'battle' would be the right thing to say), that is to say Healfdene and Eywysl, and Inwær* [his tyranny relinquished] also hastened to the hall of the infernal one, and so did senior chiefs of theirs, both jarls and other noblemen.)

It appears that early accounts reported this event differently. This would explain the considerable confusion that arose in later manuscript compilations. The aforementioned *Chronicle of John of Worcester*, for example, drawn from various sources in the late 11th and 12th centuries, tells of two separate battles, one at Tettenhall and the other at *Wodnesfeld* when in fact the two sites are only a few miles apart and both names were used to describe the same field of battle (Darlington et al. 1995, p. 365, note 8).

MSS BC also report on Tettenhall twice, but this is because the text of the *Mercian Register*, which included its own brief account of the conflict, was inserted as a complete block of annals after the main Chronicle entry for 914. The scribe of MSS D apparently tried, with little success, to merge his sources and ended up reporting the battle three times (Wainwright 1945, p. 385; Whitelock 1979, p. 210, note 5).

In MSS E, whose compiler had access to now lost northern sources, a short entry (one of very few in this version for this period) tells of a single battle *æt Teotanheale*. However, the compiler describes the English army, most unusually, as a *here*, a term normally associated with 'devastation and robbery' otherwise used almost exclusively to denote Scandinavian enemy forces (Irvine 2004, p. 54; Bosworth and Toller 1898, p. 532). All these reports, however, refer to a single battle fought at Tettenhall in 910 from which the English emerged victorious and where many Scandinavians died, including two or (given Æthelweard's version) possibly three kings.

The kings associated with the battle of Tettenhall (Healfdene, *Auisle* and *Ívarr*) are significant for the purposes of this discussion. Who were they, and where did they rule? David Dumville (2005, p. 44) observed that their names are the same as 'the names of the three Viking brothers from Dublin who were active in the 860s and 870s (*Auisle* was killed in 867), and they are no doubt here those of their children or grandchildren.'

In the same paper, Dumville argued from Irish (and other) sources that the OI *Dubgaill* (black foreigners) as opposed to the *Finngaill* (fair or white foreigners) ruled over both Dublin and York in the 9th and 10th centuries, and that the kings killed at Tettenhall were therefore members of the *Dubgaill* dynasty based at York (Dumville 2005, pp. 83–85; Downham 2007, pp. xvi–xx, 14–15, 22–23; 2008, p. 344). Alfred Smyth (1975, p. 75) also placed these kings at York, writing, '[w]e know nothing of the York rulers from the death of King Knútr in *c.* 902 until the mention of two Danish kings, *Eowils* and *Healfdene*, who were slain, according to the Anglo-Saxon Chronicle, in a battle at Tettenhall.'

It should be noted, however, that no evidence of any kind, numismatic or otherwise, has ever been found to link these kings to York in this period. In a significant study, Colmán Etchingham (2014, p. 37) again reviewed the use of the *Dubgaill* and *Finngaill* terms in contemporary sources and came to a very different conclusion about their meaning. He writes,

> *Pace* Dumville and Downham, then, we may conclude that the terms *Dubgenti* [Black Heathen] and *Dubgaill* [Black Foreigners] are not used of the Dublin leadership in the era of Amlaíb [Áleifr] and Ímar [Ívarr] and their successors, except in the context of sway over York, in the earlier 10th century. Otherwise, this terminology most commonly denotes Vikings active in Britain, who were primarily Danes and who intervened in Ireland in 851–2 and 875–7. On the latter occasion, the intervention was plainly hostile to the regime of Amlaíb's and Ímar's sons.

---

* *Iguuar, Ímar,* and *Inwær* appear as variations of ON *Ívarr* in the different sources.

> When *Dubgall* occurs in the royal titles of leaders of the 10th century Vikings, it seems to denote not their Dublin-based hegemony, but their command of Viking forces outside Ireland, primarily Danish Vikings in Britain, and their influence at York.

Regarding the term for fair or white foreigners, Etchingham wrote, '[w]hen *Finngaill* re-appears in the titles of kings descended from Ímar, in the 10th century, it seems to distinguish the Dublin-based and predominantly Norwegian portion of their domain from the York–Northumbrian and predominantly Danish portion.' If this is correct, it then follows that Healfdene, *Auisle*, and *Ívarr* are likely to have been descendants of what (until its expulsion *c.* 902) had been a Dublin-based dynasty of primarily Norwegian provenance unrelated to and often in conflict with the Danish rulers based at York.

According to this interpretation, not until another 'grandson of Ívarr,' the Hiberno-Norse King Ragnald, captured York in 919 (MSS DE) and began his rule in that city was a York–Dublin axis established. This, then, would explain why no evidence exists to link the kings slain at Tettenhall to York for they never ruled over that city or its territories, but, instead, over lands in the Wirral, Lancashire, Cumbria, or other regions in and around the Irish Sea.

Possible evidence that Scandinavian kings were once active in north-west England may be suggested by the place-name of Coniston in Cumbria. Eilert Ekwall (1922, p. 215) observed that *Conigeston*, the 'king's estate or village,' may derive from West Norse *konungr* and refer to a 'small Scandinavian mountain kingdom.' That no coins were discovered for Healfdene, *Auisle*, and *Ívarr* is also significant, for Hiberno-Norse rulers did not mint their own coinage at that time (Williams 2007, p. 202; Edwards 1998, p. 57; Sheehan, 2001, p. 53). The first of their number known to have done so (incorporating what appear to be pagan symbols into the designs) is the aforementioned Ragnald, shortly after he took control of territories formerly under Danish rule in eastern England, at some point in the second decade of the 10th century.

Scandinavian Holds of ASC

Also of particular interest for this discussion is a title given to certain Viking leaders included in the 910 list (BCD) of kings and *eorls* killed at Tettenhall. The title is *hold* which, in Old English texts, is used only in reference to Scandinavians. Such leaders were, according to Norwegian law texts, free men in possession of allodial lands (Larson 1935, p. 12).

Hold appears to be a loan-word in Old English from Old Norse (Fritzner et al. 1886, pp. 288–290; Björkman 1900–1902, pp. 283 and 290, note 1; 1901, p. 7; Hofmann 1955, §§ 205, 222, 235).* If this is correct, it may offer further support for a primarily Norwegian (West Norse) Irish Sea Viking interpretation of the evidence relating to the Scandinavian enemy in 909 and 910. According to Gillian Fellows-Jensen (1989, p. 90) the *hold* term is 'considered to have developed from a Germanic stem *haluþ-*, in which the *u* has caused mutation of the *a* to *ǫ* (*au*) in the recorded Norwegian forms.' It is, in fact, only found in Scandinavian texts in West Norse (Norwegian), the language from which the etymological evidence would seem to suggest that it entered Old English. Fellows-Jensen found this rather surprising:

> The Old Danish form of the word would have been *hald*, which would have been expected to have been borrowed into English as *hald*, much as *fatu*, which developed into *fǫt* in West Scandinavian, was borrowed into English as *fatu*…. If the Old English word *hold* really is a loan-word from Scandinavian, then the *o* can only be explained as an orthographic substitution for mutated *ǫ*. The possibility should perhaps be taken into account, however, that the spelling *hold* reflects the influence of the etymologically related adjective Old English *hold*, Old Danish *hold*, *huld* 'faithful, loyal'…. Association with this word may well have influenced not only the form taken by the loan-word but also its content, for the word *hold* in England was used of men of much higher status than the *höldar* in Norway. *Hold*s are named in company with Danish kings and jarls in the Anglo-Saxon Chronicle, while in a list of wergelds dating from the 11th century the *hold* is ranked with the king's high reeve as having half the wergeld of a bishop or ealdorman and twice that of a thegn.

---

* Richard Dance (2003, pp. 152–153, note 135) expressed some reservations to this view.

That Fellows-Jensen felt the need to formulate an alternative explanation for how the *hold* term may have developed in Old English other than as a loan-word from West Norse seems to derive from her adherence to the traditional interpretation of early 10th insular history in which *hold*s were likely to have been Danes for they 'are named in company with Danish kings and jarls in the Anglo-Saxon Chronicle.' But were these *hold*s and the kings and jarls they accompanied at Tettenhall necessarily Danes? Could it not be that the *hold* term is, instead, further indication of primarily Norwegian, Irish Sea Viking involvement in north-west England in the early 10th century which was not specifically reported in the Chronicle?

An important early attestation of the term *hold* in Old Norse is found in *Eiríksmál*, an anonymous and incomplete skaldic poem in praise of the late *Eiríkr blóðøx* (Eric Bloodaxe), the king of Northumbria. It is traditionally held to have been commissioned by his queen, Gunnhild, following his death on Stainmore in 954 and may date to as early as the mid 10th century.

The poem refers to several men of high standing, *hold*s, who accompanied *Eiríkr* into *Valhöll* (Einarsson 1985, pp. 77–78; Finlay 2004, p. 58; Seeberg 1978, p. 106). Another direct, though considerably later, link between *hold*s and Northumbria comes from an 11th century list of wergelds mentioned by Fellows-Jensen. Known as the 'Law of the Northumbrian Priests,' it pertains to 'Northumbria and the territory of the Five Boroughs' and indicates that the position of *hold* continued in that region for several generations beyond the early 10th century (Fellows-Jensen 1989, p. 89).

The most economical explanation, it would seem, for how the *hold* term came into Old English is for such leaders, whatever their specific function, to have been of primarily West Norse or Norwegian background. Indeed, it could be argued that *hold* is likely to be a loan-word from West Norse because the *hold*s of the Chronicle are named in company with (predominantly) Norwegian Irish Sea Viking kings—the 'sons' or 'grandsons of Ívarr.' Regarding Fellows-Jensen's observation that *hold*s in England seem to have enjoyed higher status than *höldar* in Norway, it is hardly surprising should it be found that various ranks and positions of authority such as *hold*s developed differently over time and in different kingdoms or kingships. The fact that *hold*s in England may have exercised greater authority than the *höldar* of Norway does not rule out the possibility that the *hold*s of England were largely made up of Irish Sea Vikings speaking a West Norse dialect. It is also significant that the term never appears in Danish or Swedish literary contexts.

In what may be seen as further possible support for this hypothesis, Fellows-Jensen's 1989 article also considered possible place-name evidence in Britain for the *hold* term. Holderness, a peninsula on England's east coast between the North Sea and the Humber appears to have *höldr* as its specific. Fellows-Jensen (1989, p. 90) noted that, '[i]f the district of Holderness really did take its name from the *hold* who wielded authority over it, then this too would support the view that the hold in England can have been no mere freeman with rights over a small portion of allodial land.'

There is another possible although admittedly tenuous link of a place-name, a major land-holding under the authority of a Scandinavian *hold*, and annal 910 in ASC. It has to do with the district of Amounderness, a large territory in Lancashire (Ekwall 1922, p. 139). In a grant dated to the 930s, King Athelstan made a gift of Amounderness to the church of York (Sawyer 1968, p. 407). The authenticity of this grant has been questioned by some, though Fellows-Jensen (1989, p. 88) regarded it as 'a genuine document that has been somewhat corrupted in transmission.'

Whitelock also regarded it as genuine (1979, pp. 548–550). The place-name's specific is derived from ON *Agmundr*, a personal name shared with one of the *hold*s killed at Tettenhall (Fellows-Jensen 1992, p. 40; 1998, p. 18). It is possible (although impossible to verify) that *Agmundr* of Amounderness and *Agmund hold* at Tettenhall may have been one and the same (Bugge 1904, p. 317; Fellows-Jensen 1989, p. 90; Higham 1992, p. 24). Wainwright (1975, p. 223) noted that, 'if the identification could be accepted as proved it would be of considerable historical importance, for it would show the Lancashire Norsemen raiding into southern England and would imply a greater degree of military organization than we have allowed.'

Evidence for Scandinavian Presence in North-West England

The evidence for a large-scale Scandinavian migration into north-west England beginning in the early 10th century is considerable. One of the main challenges facing those who seek to construct a coherent narrative from that evidence, however, is the near complete dearth of contemporary written sources for the region. Alex Woolf (2007, p. 132) noted, for example, that 'the western coast of Northumbria from the Mersey to Ayrshire is a "chronicle blind-spot" in the early middle ages and almost no events occurring there are recorded.'

Related Textual Evidence

One event that took place across the Irish Sea and had a significant bearing on Viking presence in the north-west at the time is reported in the *Annals of Ulster*. It states that in 902 a coalition of Irish kings turned against the Vikings whose seat of power had for decades been centred on Dublin. Another source, known as the *Fragmentary Annals of Ireland*, tells of the adventures of *Ingimundr*, a Viking leader who survived the Dublin expulsion.

Two other passages suggesting Viking activity in the north-west then are found in *Historia de Sancto Cuthberto*, a Latin work compiled in the mid 11th century that may contain texts dating to as early as the mid 10th century. Two Anglo-Saxon dignitaries are reported as having left north-west England by 918 at the very latest, and quite possibly several years earlier (Johnson-South 2002, p. 61). One of these, Tilred, was abbot of Heversham in Westmorland. For reasons not explained in the text, he left Westmorland and headed east where he purchased the abbacy of Norham on Tweed from Bishop Cutheard (918 or possibly 915).

The other dignitary is Alfred, son of Brihtwulf, who,'fleeing from pirates, came over the mountains in the west and sought the mercy of St. Cuthbert and bishop Cutheard so that they might present him with some lands.' It seems reasonable to assume that, as with Tilred, Alfred's lands would have come at a price to satisfy the good bishop.

From these short passages (see also Griffiths, Chapter 2) it seems that some disruption (certainly in the case of Alfred at least) caused by Vikings in the early part of the reign of Edward the Elder led to the departure of these men (and quite possibly others) from the north-west. Significantly, however, they appear to have been in possession of sufficient funds to resettle comfortably in their new surroundings.

Place-Names

Another hindrance to piecing together the story behind the Scandinavians' arrival in north-west England lies in the uncertainty related to dating much of the evidence. Place-names, for example, began to appear in written records from the 11th century and many were not recorded until much later. Among the few significant exceptions is the grant of Amounderness, as noted above, in a charter from the 930s. Here is an example of an entire administrative district in Lancashire bearing the name of what must have been a Scandinavian overlord before 934, suggesting that elite Viking leaders controlled major seats of power in this region at that time.

Angus Winchester (1985, p. 99) observed that 'the major territorial divisions down the Irish Sea coast of north-west England—Allerdale, Copeland, Furness, Amounderness—all bear names of Scandinavian origin.' Furthermore, the sheer abundance of ON place-names in parts of north-west England (Wainwright 1975, pp. 181–279; Fellows-Jensen 1985; Whaley 1996) suggests the presence of very large numbers of Scandinavians whose influence on the spoken language was both considerable and long-lasting.

Such influences can also be traced in local dialects (Ellis 1985, pp. 161–167), the occasional runic inscriptions from later centuries such as at Carlisle and Pennington in Cumbria (Barnes and Page 2006, p. 19), and also in an early text about territories in Cumbria and known as *Gospatric's Writ* (Harmer 1952, pp. 419–424). Some scholars have noted a West Norse or Norwegian bias in the place-names of north-west England (Smyth 1975, p. 77; Whaley 1996, p. xxi).

Coin and Silver Hoards

Two types of evidence relating to Scandinavian presence in north-west England allow for closer dating— one quite exact and the other to within a few decades. The first and more precise items are coin hoards which, if they contain sufficient coins for comparison, can often be dated accurately to within a few years of their deposition. Large numbers of silver and coin hoards have been found in north-west England. Two of these have particular significance for they may support the theory of a possible link between north-west England and the events recorded in ASC for 909 and 910.

The first hoard was found near Harkirk in Lancashire (north-west of Liverpool) in 1611. The coins have not survived but an engraving made near the time of discovery allowed numismatists to determine that the hoard included pennies from at least four series of Edward's coins minted *ca.* 910, all bearing the names of moneyers from south-east England (Edwards 1998, pp. 42–45; Graham-Campbell 2001, pp. 219–220). The second hoard discovered in Shrewsbury in 1936 contained similar Anglo-Saxon issues (Robinson 1983, pp. 7–13).

The dates of the coins, their provenance, and places of deposition led Stewart Lyon (2001, p. 75) to suggest that '[i]t is conceivable that both these hoards are related to Edward's campaign of 909–910 against the Northumbrian Danes, which ended in the battle of Tettenhall, less than 30 miles east of Shrewsbury.' This accords well with the hypothesis that the Scandinavians of the *norðhere* in 909 and 910 were, in fact, Irish Sea Vikings who, following the destruction of Dublin as a Viking seat of power in 902, re-established themselves in north-west England.

Stone Sculpture

Over the past few decades, significant advances have been made in the study of stone sculpture in Britain. This second type of evidence, after coins, is susceptible to closer dating. This is important for, as B. J. N. Edwards noted, '[i]n a distribution map of Viking Age artefacts from North West England '… the largest number of symbols relates to sculpture' (1998, p. 69).

Research into the possibility that Scandinavian stone sculptures in Britain may have functioned as symbols relating to the administration of secular power has been published in recent years (Stocker 2000; Stocker and Everson 2001; Sidebottom 2000). Here again is evidence suggesting the presence of elite Scandinavians with Hiberno-Norse tastes who were more than mere settlers and possibly ruled over parts of the north-west in the first half of the 10th century.

Archaeological Record

Although burials are, as a rule, less easily dated than coin hoards or stone sculpture, the window for the creation of furnished graves in north-west England was narrow and probably did not extend much beyond *ca.* 950. Whilst the number of such graves is not great, compared to the rest of England, it is proportionally high. Richly furnished burials, particularly high-status graves under mounds, seem unlikely to represent mere personal taste or family preference (Andrén 1993; Skre 1997). Rather, when considered with stone sculpture monuments, they suggest that Scandinavians with considerable political influence were involved in this region in the first half of the 10th century.

Summary

The foregoing was not intended as a comprehensive survey, but simply as a brief introduction to the types of evidence available for study. It was on the basis of such evidence that Wainwright (1975, p. 226) wrote, 'It may be stated with some confidence that the hordes of Norsemen in northern England and the alarm they inspired there provide the key to understanding of otherwise inexplicable developments. It might not be too much to say that the history of England in the early 10th century turns largely upon the Norsemen, upon what action they took and, above all, upon what action they might have taken.'

Wainwright's 'key to understanding' may be more significant than even he seems to have anticipated, but first we should consider the origins of the received interpretation.

Post-Conquest Textual Evidence for the Battle

When the Anglo-Normans began the challenging task of gathering source materials and compiling their historical works in the late 11th and 12th centuries, hoards such as those at Harkirk, Shrewsbury, and

Cuerdale remained hidden in the earth, as did the pagan Scandinavian burials at Aspatria, Hesket-in-the-Forest, Cumwhitton, Claughton, and elsewhere.

The significance of the magnificent stone cross and other sculptures at Gosforth, the hogbacks at Lowther, and all the other Anglo-Scandinavian stone monuments in north-west England had, in all likelihood, been forgotten by the late 12th century, even to the few who were aware of their existence. 'It is a curious fact,' wrote Wainwright (1975, p. 133), 'that for nearly a thousand years the Norse immigration into north-west England lay outside the knowledge of historical writers. This was due in the main to the lack of literary record, to the silence of the chroniclers.' This caution should be borne in mind when reading of the events of 909 and 910 in the works of the Anglo-Normans.

Henry of Huntingdon

Henry of Huntingdon's *Historia Anglorum* was compiled in the first half of the 12th century and enjoyed wide circulation in later centuries (Whitelock 1979, pp. 118–119). Henry's version of the events of 909 and 910 reads as follows in translation (Greenway 1996, pp. 301 and 303):

> In his fifth year, King Edward made a truce with the East Anglians and the Northumbrians at Tiddingford. In the following year, the king sent a strong army from Wessex and Mercia, which took enormous plunder from the army that was in Northumbria, both in men and cattle. They killed many of the Danes and continued to ravage their territory for five weeks. In the following year, the Danish army came to ravage Mercia. The king gathered together a hundred ships and sent them against the army. When they appeared, the army thought they would receive aid and would be able to proceed more safely wherever they wished. Next, the king sent an army after them from Wessex and Mercia, which followed them from the rear as they turned back homewards, and attacked them. A huge battle commenced. But the Lord struck down the heathens in a great defeat, and bloody death consumed many thousands of them. Their leaders fell to the ground in confusion and were destroyed and made filthy in the dust.

Although taken from ASC, Henry's account introduces a chronological error into the text. Having correctly dated the Tiddingford treaty to the fifth year of Edward's reign (*ca.* 905), Henry then states that the English raid (conducted against the *norðhere*) took place '[i]n the following year.' This establishes a relationship between these events which is not specified in ASC's version of the event. Henry also appears to have struggled with interpreting this passage.

According to ASC 910, it was because the Scandinavians thought that 'the greater part of [Edward's] forces was on the ships' off the coast of Kent that they assumed 'they could go unopposed wherever they wished' (Whitelock et al. 1961, p. 61). Henry states, '[w]hen they appeared, the army thought they would receive aid from them, and would be able to proceed more safely wherever they wished.' This demonstrates that substantial misinterpretations of original sources could be and indeed were introduced by later compilers (Greenway, 301, note 96). Whilst York is not named in Henry's account, the link between the Northumbrians at Tiddingford in 905 and 906 and the English raid 'the following year' against the 'army in Northumbria' clearly implies that Danes at York were the targets of the 909 attacks.

John of Worcester

Formerly known to historians as 'Florence of Worcester,' it is now generally acknowledged that 'a monk named John' played the major role in compiling the *Chronicon ex Chronicis* (Darlington et al. 1995, p. xviii). John worked on his *Chronicon* in the decades before Henry's *Historia Anglorum*. His account of the 909 and 910 events reads in translation (Darlington et al. 1995, pp. 363, 365, and 367):

> 909: The bones of St. Oswald, king and martyr, were brought from Bardney to Mercia. The invincible king Edward, because the Danes violated the pact which they had made with him, sent an army of West Saxons and Mercians into Northumbria. When they came there, they ravaged the country without ceasing for almost 40 days, killing very many Danes. They also brought

back with them many captives, and very great booty, and the Danes' kings and nobles were compelled, willy-nilly, to restore the peace they had broken with King Edward, that is, the Elder.

910: In Staffordshire, in a place called Tettenhall, a famous battle took place between the English and the Danes, but the English were able to gain the victory. In the same year, the victorious King Edward the Elder, assembling a hundred ships, chose soldiers, ordered them to embark, and commanded them to meet him in Kent, whither he would travel by land. Meanwhile, the army of the Danes who had settled in Northumbria again broke the treaty which they had established with him and, rejecting entirely the justice which he and his men had offered them, laid waste the lands of Mercia with boldness and daring, thinking no doubt that his greater strength was at sea, and that they could go wherever they wished without a battle. When the king heard of this, he sent a combined army of West Saxons and Mercians to repel them. They, when they engaged the Danes in a field which is called Woden's field in English, as they were returning from their work of destruction, slew two of their kings, Eowils and Healfdene, brother(s) of King Inguar, and two earls, Ohter and Scurfa, and nine of the greater nobles and many thousands of the men, and when they had put the rest to flight they recaptured all the booty.

John reports on the battle of Tettenhall twice, thinking perhaps that Tettenhall and *Wodnesfeld* refer to two different engagements. Although John maintains a correct chronology between the Tiddingford treaty and the battle of Tettenhall, his claim that it was 'because the Danes violated the pact with him' that Edward had sent his men on their five-week raid in 909 is a clear reference to the Tiddingford treaty.

John also writes that the English ravaged 'Northumbria' rather than the 'northern army' of ASC (Darlington et al. 1995, p. 363, note 3). In his second account of Tettenhall in annal 910, John describes the enemy as *Danorum exercitus qui Norðanymbriam inhabitabant* (the army of the Danes who had settled in Northumbria), thus eliminating the ambiguity in ASC's version about the identity of this Scandinavian force.

William of Malmesbury

William of Malmesbury's *Gesta Regum Anglorum*, compiled in the first half of the 12th century, was also widely read throughout the Middle Ages. William often synthesised the content of his sources into his own narrative prose. For example, when recounting the various achievements of Edward the Elder, the whole of the king's reign was condensed into the following laudatory summation (Mynors et al. 1998, p. 197):

In the year of our Lord 901, Alfred's son Edward ascended the throne, which he held for 23 years. He was much inferior to his father in book learning, but in his power and glory as a king there was no comparison, for one united the two kingdoms of the Mercians and West Saxons, holding the former in name only, because it was entrusted to Ealdorman Æthelred; the other first took full control of the Mercians after Æthelred's death; then he defeated in battle and subjected the West and East Angles and the Northumbrians, who had already grown into one nation with the Danes, the Scots who dwell in the northern part of the island, and the Britons (whom we call Welsh); nor did he come off second best in any contest.

William's statement that Edward 'defeated in battle and subjected the West and East Angles and the Northumbrians, who had already grown into one nation with the Danes,' like Henry of Huntingdon and John of Worcester, goes beyond what is stated in the Chronicle by placing all these events in the Danish-ruled kingdoms of eastern England.

Given the influence these Anglo-Norman authors' writings exerted for centuries, it would have been difficult for later scholars to interpret annals 909 and 910 in any way other than to assume that the *norðhere* defeated by the English at Tettenhall had been the Northumbrian army with its centre of power at York.[*]

---

[*] According to Darlington et al. (1995, p. xvii) the *Chronicle of John of Worcester* was 'well known and frequently used in the 12th and 13th centuries; and it has long been regarded by modern scholars as of value for our knowledge of the Anglo-Saxon period.'

Modern Scholarship and Battle of Tettenhall

In his four-volume *History of the Anglo-Saxons*, first published between 1799 and 1805, Sharon Turner (Loyn, 2004) quoted extensively from the Anglo-Normans. Having described the Tiddingford treaty as a 'peace' which, two years after the battle of the Holme, 'restored amity between the Anglo-Saxons and Anglo-Danes,' Turner (1852, pp. 144–145) gave this account of the 909 and 910 conflicts, which he related to the 905–906 treaty at Tiddingford:

> But war was soon renewed between the rival powers. With his Mercians and West Saxons, Edward, in a five weeks' depredation of Northumbria, destroyed and plundered extensively. In the next year, the Northerns devastated Mercia. A misconception of the Danes brought them within the reach of the king's sword. While he was tarrying in Kent, he collected 100 ships, which he sent to guard the south-eastern coast, probably to prevent new invasions. The Danes, fancying the great body of his forces to be on the seas, advanced into the country to the Avon, and plundered without apprehension, and passed onwards to the Severn. Edward immediately sent a powerful army to attack them; his orders were obeyed. The Northerns were surprised into a fixed battle at Wodensfield, and were defeated, with the slaughter of many thousands. Two of their kings fell, brothers of the celebrated Ingwar, and therefore children of Ragnar Lodbrog, and many earls and officers.

Turner's conclusion that after the Tiddingford treaty 'war was soon resumed between the rival powers' (thus linking ca. 905 treaty with the 909 raid) was the only interpretation he could have made of these texts. Early 19th century scholars, like their 12th century counterparts, could not have conceived of a Scandinavian 'power' to rival the Anglo-Saxons in 909 other than those known from ASC to have settled in eastern England. As for Turner's influence on Anglo-Saxon historiography, Philip Grierson (1945, p. 247) remarked that 'Turner's work may now be forgotten by most historians, but it occupies an honourable place in the history of Anglo-Saxon scholarship. Even in the last century it was never fully superseded.'

20th Century

Despite the growing awareness from the 1850s of early 10th century Scandinavian settlement in England, interpretations by scholars in the 20th century of the events in Britain of 909 and 910 strayed little from the traditions established in previous generations. In his *A History of the Vikings*, T. D. Kendrick (1930, pp. 246–247) concluded that the English victory at Tettenhall resulted in 'a complete overthrow of the Danes' and led to 'a new phase in the history of the Danelaw, for it was no longer Mercia and Wessex that lived in fear of attack. It was the Danes now that were in peril.'

Similarly, Stenton (1971, pp. 323–334), in his monumental *Anglo-Saxon England*, the first edition of which was published in 1943, stated that the Danish Northumbrian kingdom had been so devastated by the defeat at Tettenhall as to have been rendered powerless to repel Edward and Æthelflæd's 'gradual reduction of East Anglia and the Danish midlands.' This conclusion was made in part by observing that 'the chronicles which describe the West Saxon advance against the southern Danes give no hint of Northumbrian intervention.' The argument, therefore, is based on what the Chronicle does *not* say.

Dorothy Whitelock (1979, p. 33) agreed with this conclusion and in the introduction to her oft-cited *English Historical Documents* (first published in 1955) claimed that the peace of Tiddingford had been 'of short duration, and it was not until after the great defeat of the Northumbrians at Tettenhall in 910, where they lost two kings and many high officials, that Edward and his sister Æthelflæd, lady of the Mercians, were able to begin their great concerted action to reconquer the Danelaw.' This interpretation of events undergirds the translation Whitelock co-edited, as is seen in the annal for 910 (1961, p. 62):

> In this year the army in Northumbria broke the peace, and scorned every privilege that King Edward and his councillors offered them. And the king had collected about 100 ships, and was

then in Kent, and the ships were on their way south-east by sea towards him. Then the Danish army thought that the greater part of his forces was on the ships, and that they could go unopposed wherever they wished. Then the king learnt that they had gone on a raid. He then sent his army both from the West Saxons and Mercians, and they overtook the Danish army when it was on its way home and fought against it and put the army to flight and killed many thousands of its men.

Here MS A's phrase 'the army in Northumbria broke the peace' appears common to all versions, for there is no indication that A is at variance with the other manuscripts stating that it was 'the peace in Northumbria' that had been broken by *se here*.[*] Then, two sentences later, *se here* (the army) is translated as 'the Danish army,' and again in the following sentence. As a result, there appears to be a progression from Edward's treaty at Tiddingford *ca.* 905 with 'the East Angles and Northumbrians' to the 909 raid against the 'northern army' in which 'those Danes' were killed, and then to 'the army in Northumbria' (described twice again as 'the Danish army') breaking the peace in 910. The reader is left, therefore, with the impression that the Scandinavians involved were the Northumbrian Danes of York.

Conclusion

The received interpretation concerning the battle of Tettenhall rests upon assumptions that are open for discussion. For example, the political relationships of the Anglo-Saxons, Northumbrians, East Anglians, and Irish Sea Vikings in the early 10th century are unknown, and it should not be assumed that Scandinavians (however that identity may have been understood or expressed) regularly fell in together against the English.

Perhaps in the aftermath of the debacle into which Æthelwold led them (that ended so dismally for them at the battle of the Holme *ca.* 902), the Scandinavian rulers of East Anglia and Northumbria allied themselves with Edward at Tiddingford in 905 or 906 in response to the threat of a rapidly expanding Irish Sea Viking presence in north-west England.[†]

If this was the reason for the Tiddingford treaty, it could explain why Edward according to MS E agreed to the treaty *for neode* (by necessity) as an alliance to help secure his north-western borders against a new, powerful, and dangerous enemy.

While the exodus of refugees from Dublin in 902 would not alone have amounted to a mass migration of Scandinavians into north-west England, the significance of that expulsion may lie not in the numbers but in the status of the exiles. It was no insignificant settlement of unwelcome foreigners who were driven from Ireland, but a great seat of power and the dynasty that went with it.

The facts that (1) Amounderness in Lancashire bears the name of a Scandinavian and was purchased by Athelstan in the 930s 'from hateful servitude' (Whitelock 1979, p. 547), (2) the Cuerdale hoard (deposited *ca.* 905-10) is a Viking treasure with clear links to Ireland that signifies immense wealth and status,

---

[*] Michael Swanton's translation of annal 910 (2000, p. 97) also states 'the raiding-army in Northumbria broke the peace.'

[†] The Cuerdale hoard (deposited *ca.* 905-10 and discovered in 1840) is an interesting find to consider in this context. The largest part of its coinage (some 5000 of over 7000 coins) comes from the Danish-ruled kingdoms of Northumbria and East Anglia. These coins are closely die-linked and many appear as if fresh from the mint (Lyon, 1961, pp. 109–110; Graham-Campbell, 2001, pp. 220–222). While the Carolingian coins and some of the assorted silver pieces that constitute a significantly smaller part of the hoard are commonly regarded as Viking loot, it is often assumed that the Northumbrian and East Anglian coins symbolise solidarity and cooperation in some form between the Danish-ruled kingdoms of eastern England and the Irish Sea Vikings. The hoard perhaps was an 'army pay-chest' intended to reverse the Dublin expulsion (Grierson et al., 1986, pp. 320–321; Graham-Campbell, 1987, pp. 329–330; Downham, 2007, pp. 28, and 78–79). As an alternative to this interpretation (and assuming that the Dublin and York leaderships were enemies rather than allies), these coins may represent instead tribute taken from the Danes of Northumbria and East Anglia by the Irish Sea Vikings after their expulsion from Dublin and Ireland. Given the time proximity between the estimated date for the deposition of the hoard at Cuerdale on the Ribble (*ca.* 905) and Edward's *frið* with the Northumbrians and East Anglians at Tiddingford in the south-east (*ca.* 905–906), it makes for interesting conjecture that this treaty may represent a pact between the Anglo-Saxons and Danelaw rulers against the newly-established Irish Sea Vikings in north-west England. Should the Cuerdale hoard represent extortion on an unprecedented scale, it might explain the apparent economic disaster some numismatists have suggested may lie behind the sudden debasement in Danelaw coinage after 905 (Blackburn, 2004, pp. 333 and 340).

and (3) Tilred and Alfred (fleeing pirates) could have, before 918, left the north-west and afforded to re-establish themselves in eastern England, may with other evidence be seen to support the theory that at least some of the displaced Dublin dynasty leaders bought their way into power in north-west England after 902 rather than attempting a land take through violence (Winchester 1985, 98–99).*

In what seems an interesting coincidence, we know from later charter evidence that Edward and Æthelred ordered their men to buy back land 'from the pagans' (Sawyer 1968, S396–S397) before Æthelred's death in 911. Also, there seems to have been a complete hiatus in the issue of land grants from *ca.* 909 to 924 (909 was the same year Edward sent out his men against the northern army), possibly reflecting a deliberate policy enforced by Edward to the end of his reign (Keynes 2001, pp. 55–56).

If the leaders of the Dublin dynasty took control of large parts of north-west England by legal means after their expulsion from Ireland, there may have been little to gain from altering the basic structures of that society. Ownership of their new lands, for example, if purchased, would have been indisputable. This may explain why many of the existing place names remained unchanged (Fellows-Jensen, 1998, p. 20). Also, though the land holders who fled would undoubtedly have taken their families and top retainers away with them, the greater part of the existing population may have been left to survive as best they could under new overlords.

If this is correct then, despite signs of considerable continuity in the region's administrative structures across many generations, in a sense everything changed for much of north-west England would have come under the direct rule of the sons and grandsons of *Ívarr*. This in turn could have opened the way for Scandinavian immigration and settlement on a very large scale (without the need for violent displacement) in the following decades under the patronage and to the advantage of the Dublin dynasty rulers.

It was argued above that the *Denisc*, *norðhere*, and *Norðanhymbre* terms in the ASC are not specific in their descriptions of the Scandinavian army attacked by the English in 909 or the nature of the peace the enemy violated in 910. Two aspects of annal 910: (1) a possible link between the names of the kings killed at Tettenhall and the dynasty been based at Dublin until 902 and (2) the West Norse (Norwegian) origin of the OE *hold* term were suggested as indicating a primarily Norwegian Irish Sea Viking leadership for the *norðhere* in 909.

Furthermore, the *Mercian Register*'s report that Chester was 'restored' in 907 suggests a disturbance of some kind in the region at that time. Beyond the texts of the Chronicle, the following points were also considered: (1) whether the notice in Æthelweard's *Chronicon* concerning Æthelred's rule over parts of Northumbria and his role in the 'peace of Northumbria' broken by the Scandinavians in 910 may have involved Northumbrian territories west of the Pennines; (2) the lack of numismatic or other evidence to link the kings killed at Tettenhall to the Danish kingdom of Northumbria based at York; (3) the Harkirke and Shrewsbury hoards that included pennies of Edward the Elder minted in Wessex *ca.* 910 and may thus establish a link between north-west England and the events of 909 and 910; and (4) the story of *Ingimundr* as suggesting that the arrival of Hiberno-Norse settlers brought considerable disruption to the region. While any single piece of this evidence may be dismissed as inconclusive, when taken together it is tempting to see all these pieces as tipping the balance in favour of a north-west England–Irish Sea Viking interpretation for the Scandinavian enemy in 909 and 910.

If the Dublin dynasty leaders did begin to regroup, consolidating their power in the north-west after 902, rebuilding their armies and raiding the surrounding territories, such developments could easily have caused Edward sufficient alarm to send his army on a five-week raid against that 'northern army' and its territories in 909. And should this be deemed a viable interpretation of the evidence, it may be possible to consider a historical paradigm for Britain in the early 10th century in which (as Wainwright suggested) the great influx of Irish Sea Vikings and the establishment of a new base of power in north-west England takes on a far more central role.

ACKNOWLEDGMENTS

The author would like to thank David Parsons and Colmán Etchingham for their comments on early drafts of this chapter.

---

* It is possible (but not reported) that Æthelflæd's grant of land to *Ingimundr* was made in exchange for money.

REFERENCES

Andersen, P.S. (2006). *Det siste norske landnåmet i Vesterled: Cumbria–nordvest-England. Oslo,* Unipub forlag.

Andrén, A. (1993). Doors to other worlds: Scandinavian death rituals in a Gotlandic perspective. *Journal of European Archaeology,* 1, 33–56.

Angus, W.S. (1938). The chronology of the reign of Edward the Elder. *EHR,* 53, 194–210.

Bailey, R.N. (1980). *Viking Age Sculpture in Northern England.* London, Collins.

Bailey, R.N. (1984). Irish Sea contacts in the Viking period: The sculptural evidence. In *Beretning fra tredje tvaerfaglige Vikingesymposium,* G. Fellows-Jensen and N. Lund, Eds. Københavns universitet, Copenhagen, pp. 6–36.

Bailey, R.N. (2000). Scandinavian myth on Viking-period stone sculpture in England. In *Old Norse Myths, Literature and Society: Proceedings of 11th International Saga Conference,* M. Clunies Ross, Ed. University of Sydney, pp. 2–7.

Bailey, R.N. (2006). A miniature Viking Age hogback from the Wirral. *Antiquaries Journal,* 86, 345–356.

Bailey, R.N. and Cramp, R., Eds. (1988). *Corpus of Anglo-Saxon Stone Sculpture in England: Cumberland, Westmorland and Lancashire North of the Sands.* Oxford, Oxford University Press.

Barnes, M. and Page, R.I. (2006). *The Scandinavian Runic Inscriptions of Britain.* Uppsala, Institutionen för nordiska språk.

Bately, J., Ed. (1986). The Anglo-Saxon Chronicle, A Collaborative Edition. Volume 3: MS A. Cambridge, Brewer.

Björkman, E. (1900–1902). *Scandinavian Loan-Words in Middle English.* Halle, Niemeyer.

Blackburn, M.A.S. (2004). The coinage of Scandinavian York. In *Aspects of Anglo-Scandinavian York,* R.A. Hall et al., Eds. York, Council for British Archaeology, pp. 325–3349.

Bosworth, J. and Toller, T.N., Eds. (1898). *An Anglo-Saxon Dictionary.* Oxford, Clarendon Press.

Bugge, A. (1904). *Vikingerne: billeder fra vore forfædres liv.* Copenhagen, Gyldendal.

Cavill, P., Harding, S.E., and Jesch, J. (2000). *Wirral and its Viking Heritage.* Nottingham, English Place-Name Society.

Cubbin, G.P., Ed. (1996). *ASCCE,* Vol. 6: MS D. Woodbridge, Brewer.

Dance, R. (2003). *Words Derived from Old Norse in Early Middle English: Studies in the Vocabulary of the South-West Midland Texts.* Tempe, Center for Medieval and Renaissance Studies.

Darlington, R.R., McGurk, P., and Bray, J. (1995). *The Chronicle of John of Worcester.* Oxford, Clarendon Press.

Downham, C. (2007). *Viking Kings of Britain and Ireland: The Dynasty of Ívarr to A.D. 1014.* Edinburgh, Dunedin Academic Press.

Downham, C. (2008). Vikings in England. In *The Viking World,* S. Brink and N.S. Price, Eds. London, Routledge, pp. 341–349.

Downham, C. (2009). 'Hiberno-Norwegians' and 'Anglo-Danes': Anachronistic ethnicities and Viking-Age England. *Medieval Scandinavia,* 19, 139–169.

Dumville, D.N., Ed. (2002). *Annales Cambriae, A.D. 682–954: Texts A and C in Parallel.* Cambridge, University of Cambridge.

Dumville, D.N. (2005). Old Dubliners and new Dubliners in Ireland and Britain: A Viking Age story. In *Medieval Dublin 6: Proceedings of Friends of Medieval Dublin Symposium 2004,* S. Duffy, Ed. Dublin, Four Courts Press, pp. 78–93.

Dumville, D.N. and Lapidge, M., Eds. (1985). *ASCCE,* Vol. 17: *'The Annals of St Neots' with 'Vita Prima Sancti Neoti.'* Cambridge, Brewer.

Edwards, B.J.N. (1998). *Vikings in North-West England: The Artefacts.* Lancaster, University of Lancaster.

Einarsson, B., Ed. (1985). Ágrip Af Nóregskonunga Sogum: Fagrskinna—Nóregs Konunga Tal. Reykjavík, Brill.

Ekwall, E. (1918). *Scandinavians and Celts in the North-West of England.* Lund, C.W.K. Gleerup.

Ekwall, E. (1922). *The Place-Names of Lancashire.* Manchester, Manchester University Press.

Ellis, S. (1985). Scandinavian influences on Cumbrian dialect. In *The Scandinavians in Cumbria,* J.R. Baldwin and I. Whyte, Eds. Edinburgh, Scottish Society for Northern Studies, pp. 161–167.

Etchingham, C. (2014). Names for the Vikings in Irish Annals. In *Celtic-Norse Relationships in the Irish Sea in the Middle Ages 800–1200,* J. V. Sigurðsson and T. Bolton, Eds. Leiden and Boston, Brill, pp. 23–38.

Fellows-Jensen, G. (1989). Amounderness and Holderness. In *Studia Onomastica: festskrift till Thorsten Andersson den 23 februari 1989: With English Summaries,* L. Peterson and S. Strandberg, Eds. Stockholm, Almqvist & Wiksell, pp. 57–94.

Fellows-Jensen, G. (1992). Scandinavian place-names of the Irish Sea province. In *Viking Treasure from the North West: The Cuerdale Hoard in its Context*, J. Graham-Campbell, Ed. Liverpool, National Museum Galleries Merseyside, pp. 31–42.

Fellows-Jensen, G. (1995). *The Vikings and their Victims: The Verdict of their Names.* London, Viking Society for Northern Research.

Finlay, A. (2004). *Fagrskinna: A Catalogue of the Kings of Norway.* Leiden, Brill.

Fritzner, J., Unger, C.R., and Hødnebø, F. (1886). *Ordbog over det gamle norske sprog.* Chicago, J.T. Relling.

Graham-Campbell, J.A. (1987). Some archaeological reflections on the Cuerdale Hoard. In *Coinage in Ninth-Century Northumbria: Tenth Oxford Symposium on Coinage and Monetary History*, D.M. Metcalf, Ed. Oxford, British Archaeological Reports, pp. 329–354.

Graham-Campbell, J.A. (2001). The northern hoards: from Cuerdale to Bossall/Flaxton. In *Edward the Elder 899–924*, N.J. Higham and D. Hill, Eds. London, Routledge, pp. 212–229.

Graham-Campbell, J.A. and Philpott, R. (2009). *The Huxley Viking Hoard: Scandinavian Settlement in the North West.* Liverpool, National Museums Liverpool.

Greenway, D.E., Ed. (1996). *Historia Anglorum: The History of the English People.* Oxford, Clarendon Press.

Grierson, P. (1945). Review of 'Anglo-Saxon England' by F.M. Stenton. *EHR*, 60, 247–249.

Grierson, P. and Blackburn, M.A.S., Eds. (1986). *Medieval European Coinage with a Catalogue of the Coins in the Fitzwilliam Museum, Cambridge.* Cambridge, Cambridge University Press.

Griffiths, D. (2009). The archaeological background. In *The Huxley Viking Hoard: Scandinavian Settlement in the North West*, J. Graham-Campbell and R. Philpott, Eds. Liverpool, National Museums Liverpool, pp. 13–21.

Griffiths, D. (2010). *Vikings of the Irish Sea: Conflict and Assimilation AD 790–1050.* Stroud, History Press.

Harmer, F.E. (1952). *Anglo-Saxon Writs.* Manchester, Manchester University Press.

Hart, C.R. (1982). The B text of the Anglo-Saxon Chronicle. *Journal of Medieval History* 8, 241–299.

Hart, C.R. (1983). The early section of the *Worcester Chronicle. Journal of Medieval History*, 9, 251–315.

Higham, N.J. (1992). Northumbria, Mercia, and the Irish Sea Norse, 893–926. In *Viking Treasure from the North West: The Cuerdale Hoard in its Context,* J. Graham-Campbell, Ed. Liverpool, National Museum Galleries Merseyside, pp. 21–30.

Hofmann, D. (1955). *Nordisch-Englische Lehnbeziehungen der Wikingerzeit.* Copenhagen, Bibliotheca Arnamagnæana.

Irvine, S., Ed. (2004). *ASCCE,* Vol. 7: MS E. Woodbridge, D.S. Brewer.

Kendrick, T.D. (1930). *A History of the Vikings.* London, Frank Cass.

Keynes, S. (1986). A tale of two kings: Alfred the Great and Æthelred the Unready. *Transactions of the Royal Historical Society*, 36, 195–217.

Keynes, S. (2001). *Edward, King of the Anglo-Saxons.* In *Edward the Elder*, 899–924, Higham, N.J. and Hill, D., Eds. London, Routledge.

Keynes, S. and Lapidge, M., Eds. (1983). *Alfred the Great: Asser's 'Life of King Alfred' and Other Contemporary Sources.* Penguin, Harmondsworth.

Lang, J.T. (1984). The hogback: A Viking colonial monument. *Anglo-Saxon Studies in Archaeology and History*, 3, 86–176.

Larson, L.M. (1935). *The Earliest Norwegian Laws: Being the Gulathing Law and the Frostathing Law.* New York, Columbia University Press.

Loyn, H.R. (2004). *Oxford Dictionary of National Biography.* Oxford, Oxford University Press. http://www.oxforddnb.com/view/article/27866

Lyon, C.S.S. and Stewart, B.H.I.H. (1961). The Northumbrian Viking coins in the Cuerdale hoard. In *Anglo-Saxon Coins: Studies Presented to F.M. Stenton on the Occasion of his 80th Birthday 17 May 1960*, R.H.M. Dolley, Ed. London, Methuen, pp. 96–121.

Lyon, S. (2001). The coinage of Edward the Elder. In *Edward the Elder, 899–924,* N.J. Higham and D. Hill, Eds. London, Routledge, pp. 67–78.

Mac Airt, S. and Mac Niocaill, G., Eds. (1983). *The Annals of Ulster (to AD 1131). Part I: Text and Translation.* Dublin, Dublin Institute for Advanced Studies.

Mynors, R.A.B., Thomson, R.M., and Winterbottom, M., Eds. (1998). *William of Malmesbury: Gesta Regum Anglorum.* Oxford, Clarendon Press.

Newman, R. (2009). The Cumwhitton Viking cemetery. In *The Huxley Viking Hoard: Scandinavian Settlement in the North West,* J. Graham-Campbell and R. Philpott, Eds. Liverpool, National Museums Liverpool, pp. 21–23.

O'Brien-O'Keeffe, K., Ed. (2001). *ASCCE,* Vol. 5: MS C. Woodbridge, Brewer.

Pelteret, D.A.E. (2009). An anonymous historian of Edward the Elder's reign. In *Early Medieval Studies in Memory of Patrick Wormald,* S.D. Baxter, C.E. Karkov, J. Nelson et al., Eds. Farnham, Ashgate, pp. 319–336.

Robinson, P.H. (1983). The Shrewsbury hoard (1936) of pennies of Edward the Elder. *British Numismatic Journal,* 53, 7–13.

Sawyer, P.H. (1962). *The Age of the Vikings.* London, Edward Arnold.

Sawyer, P.H. (1968). *Anglo-Saxon Charters: An Annotated List and Bibliography.* London, Royal Historical Society.

Sawyer, P.H. (1998). *Anglo-Saxon Lincolnshire.* Lincoln, History of Lincolnshire Committee for the Society for Lincolnshire History and Archaeology.

Seeberg, A. (1978). Five kings. *Saga-Book of the Viking Society for Northern Research,* 20, 106–113.

Sheehan, J. (2001). Ireland's Viking Age hoards: Sources and contacts. In *The Vikings in Ireland,* A.C. Larsen, Ed. Roskilde, Viking Ship Museum, pp. 51–59.

Sidebottom, P. (2000). Viking Age stone monuments and social identity in Derbyshire. In *Cultures in Contact: Scandinavian Settlement in England in the 9th and 10th Centuries,* D.M. Hadley and J.D. Richards, Eds. Turnhout, Brepols, pp. 213–236.

Skre, D. (1997). Haug og grav. Hva betyr gravhaugene? In *Middelaldernens symboler,* A. Christensson, E. Mundal, and E., I. Øye, Eds. Bergen, Senter for europeiske kulturstudier, pp. 37–52.

Smith, A.H. (1928). *The Place-Names of the North Riding of Yorkshire.* Cambridge, Cambridge University Press.

Smyth, A.P. (1975). *Scandinavian York and Dublin: The History and Archaeology of Two Related Viking Kingdoms,* Vol. 1. Dublin, Templekieran Press.

Stenton, F.M. (1970). *Preparatory to Anglo-Saxon England: Being the Collected Papers of Frank Merry Stenton,* D.M. Stenton, Ed. Oxford, Clarendon Press.

Stenton, F.M. (1971). *Anglo-Saxon England,* 2nd ed. Oxford, Clarendon Press.

Stocker, D.A. (2000). Monuments and merchants: Irregularities in the distribution of stone sculpture in Lincolnshire and Yorkshire in the 10th century. In *Cultures in Contact: Scandinavian Settlement in England in the Ninth and Tenth Centuries,* D.M. Hadley and J.D. Richards, Eds. Turnhout, Brepols, pp. 179–212.

Stocker, D.A. and Everson, P. (2001). Five towns' funerals: Decoding diversity in Danelaw stone sculpture. In *Vikings and the Danelaw: Select Papers from the Proceedings of 13th Viking Congress, Nottingham and York,* August 1997, J. Graham-Campbell, R. Hall, J. Jesch, et al., Eds. Oxford, Oxbow, pp. 223–243.

Swanton, M.J., Ed. (2000). *The Anglo-Saxon Chronicle,* 2nd ed. London, Phoenix.

Taylor, S. (1983). *ASCCE,* Vol. 4: MS B. Cambridge, Brewer.

Turner, S. (1852). *The History of the Anglo-Saxons from the Earliest Period to the Norman Conquest,* Vol 2. London, Longman.

Valtonen, I. (2008). *The North in the 'Old English Orosius': A Geographical Narrative in Context.* Helsinki, Societe Neophilologique.

Wainwright, F.T. (1945). The Chronology of the 'Mercian Register.' *English Historical Review* 60, 385–392.

Wainwright, F.T. (1975). *Scandinavian England,* H.P.R. Finsberg, Ed. Chichester, Phillimore.

Whaley, D. (2006). *A Dictionary of Lake District Place-Names.* Nottingham, English Place-Name Society.

Whitbread, L. (1959). Æthelweard and the Anglo-Saxon Chronicle. *English Historical Review,* 74, 577–589.

Whitelock, D. (1979). *English Historical Documents, ca. 500–1042,* 2nd ed. London, Eyre and Spottiswoode.

Whitelock, D., Douglas, and D.C.,Tucker, S.I. (1961). *The Anglo-Saxon Chronicle: A Revised Translation.* London, Eyre and Spottiswoode.

Williams, G. (2007). Kingship, Christianity and coinage: Monetary and political perspectives on silver economy in the Viking Age. In *Silver Economy in the Viking Age,* J. Graham-Campbell and G. Williams, Eds. Walnut Creek, CA: Left Coast Press, pp. 177–214.

Winchester, A. (1985). The multiple estate: A framework for the evolution of settlement in Anglo-Saxon and Scandinavian Cumbria. In *The Scandinavians in Cumbria,* J.R. Baldwin and I. Whyte, Eds. Edinburgh, Scottish Society for Northern Studies, pp. 89–101.

Woolf, A. (2007). *From Pictland to Alba: 789–1070.* Edinburgh, Edinburgh University Press.

6

The Battle of Brunanburh in 937: Battlefield Despatches

Paul Cavill

CONTENTS

ABSTRACT   The English 'Great War' of the 10th century, the battle of *Brunanburh* of 937, most likely took place at Bromborough on the Wirral. Forceful claims, however, are made for other locations and the Bromborough claim is disputed. This chapter assesses the most recent interpretations denying Bromborough's claim and supporting a location in eastern England or the Solway region. The arguments for the first are not compelling, and those in support of Burnswark in Dumfriesshire lack evidence.

Introduction

The battle of *Brunanburh* was fought by the West Saxon king Athelstan and his brother Edmund against a coalition of Scots, Strathclyde Britons, and Dublin Norsemen in the year 937, and the English won. That summary lists almost all the points of consensus that have so far been reached about the battle.

I have argued in various places that since Bromborough on the Wirral is the only place-name known to derive from Old English *Brunanburh*, it should at least be considered in any discussion of the battle site (Cavill, Harding and Jesch 2004; Cavill 2007 and 2008). But for various reasons, depending particularly on conflicting views of the military and political history of the time, widely various places have been suggested as the location of the battle, from Burnswark near Dumfries in the north to Bromswold in Huntingdonshire and neighbouring shires in the south, and from Bourne in Cambridgeshire or Lincolnshire in the east to Bromborough in the west, and many more.

A near-contemporary Old English poem about the battle appears in the *Anglo-Saxon Chronicle* and the conflict is cited in various documents of many genres from charters to verse chronicles in Old and Middle English, Latin, Anglo-Norman, Welsh, Scots, and Old Norse, throughout the Middle Ages; see Livingston (2011) in which 53 sources are collected, edited, and translated. Many aspects of these texts and traditions have been discussed, but I want to home in once more on the issue of where the battle took place, because this continues to be the source of claims and counterclaims.

The place-names are crucial in this area. I say place-names because there are in fact several different names attached to the site of the battle and its aftermath. Localising the battle with any plausibility involves dealing with this range of evidence, including the grammar, historico-linguistic changes and geographical distributions of the names and their component parts, as well as interpreting the sources as literature. Many of the suggestions about the locality of the battle either misinterpret or ignore the place-name evidence, apparently on the assumption that medieval people could not tell the difference between London and Langdon, or if they could, which they might travel to or fight at was a matter of indifference. In fact, with no maps available, place-names were more significant guides to terrain and location for travellers and armies in the Middle Ages than they are today. To dismiss or diminish the place-name evidence about a significant battle like *Brunanburh* is to eliminate what is arguably the single most important thread linking our present understanding with the historical event.

The earliest sources to name the place of the battle—the versions of the Old English poem appearing as the annal for 937 in several extant *Anglo-Saxon Chronicle* versions—give the location of the battle as *Brunanburh*, and this form or slight variations of it were recorded in the 12th century by the Latin historians Symeon of Durham (Livingston 2011, pp. 54–55; Rollason 2000), John of Worcester (Livingston 2011, pp. 56–57; Darlington and McGurk 1995–1998), Henry of Huntingdon (Livingston 2011, pp. 60–65; Greenway 1996), and many others who borrowed from or copied them (Cavill 2011, pp. 329–330).

Knowledge of the battle location was apparently soon lost, and indeed it is probable that the 12th-century historians mentioned did not know its actual location (see further below). Some theories about the location of the battle (anywhere between Devon and Northumberland) were propounded by early antiquarians (Campbell 1938, pp. 58–59, note 4; Foot 2011, p. 173). In more recent years, authoritative statements about its location or, more properly, the indeterminacy of it, have been often quoted: '... all hope of localising Brunanburh is lost,' wrote Campbell, but only the second part of the following sentence is much noted: 'Unless new evidence can be produced, an honest *nescio* is greatly to be preferred to ambitious localisations built on sand' (1938, p. 80).

Such new evidence was in large part forthcoming with publication of the survey of Cheshire place-names in the 1970s by John Dodgson (1970–1997).[*] Dodgson's survey convincingly demonstrated that Bromborough on the Wirral derives its name from an earlier *Brunanburh*, making it a clear and outstanding contender as the site of the battle. However, Dodgson was reluctant to push the Wirral claim, though he saw that if *Dingesmere*, named in the Old English poem as the area from which the defeated Norsemen fled to Dublin, could be identified nearby, it would strengthen the case enormously (Dodgson 1997, p. 263, note 11; Downham 2008, p. 104). I have published full linguistic and onomastic analyses of the ancient sources (Cavill 2008), have (with colleagues) identified *Dingesmere*, to a reasonable level of probability (Cavill, Harding and Jesch 2004; Cavill 2007), and have shown how most of the evidence converges to make Bromborough on the Wirral the most plausible location for the climactic battle (Cavill 2011).

Even since Dodgson, numerous places have been proposed and reasserted as the site of *Brunanburh*. Brinsworth (Wood 1980), Burnswark (Halloran 2005 and 2010), Bourne (Hart 1992), and Bromswold (Smyth 1987) are the front-runners amongst thirty-odd contenders, most of whose names begin with *B-*, contain *r* and often *u*. However, as Ray Page observed, 'It is hardly enough to look round for the nearest modern name beginning with *Br-* and identify that as *Brunanburh*' (1982, p. 344). Some significant problems with these localisations have been identified over the years, not least that the topographical descriptors used in the sources do not occur in the areas or apply to the topography of the places suggested (Cavill 2008), in addition to the fact that there is no evidence that these places were ever called *Brunanburh* or anything like it.

Two vigorously-proposed objections to the localisation of *Brunanburh* on the Wirral remain current and warrant further attention. One focuses on the tradition first mentioned in John of Worcester in the 12th century, that the sea-borne Viking forces entered the Humber. Michael Wood's argument, first proposed in an article (Wood 1980), reinforced in his book (1999, pp. 203–221), and renewed in public lectures, maintains that this tradition is original and accurate, and he thus locates the battle somewhere

---

[*] The survey part of Dodgson's work (Parts 1 to 4) was complete and published by 1972; the elements list (Parts 5.1(i) and 5.1(ii)) was complete and published in 1982. The introductory material and index (Part 5.2) was completed by Alexander Rumble and published in 1997.

in Yorkshire. The other argument proposed by Kevin Halloran in two articles in the *Scottish Historical Review* (2005 and 2010), seeks to dismiss the case for Bromborough and make a case for Burnswark in Dumfriesshire as the site of the encounter.* These arguments serve as the focus of the analysis below.

Humber Entry

The Humber was a hugely important waterway and boundary in Anglo-Saxon England. Old English sources mention it 38 times, not counting its inclusion in the regional name of the kingdom of Northumbria (*Dictionary of Old English Corpus*). There is a profound improbability to the idea of an invading Viking force from the west (Dublin) sailing hundreds of miles around Scotland to the east to meet another force from the west, the warriors of Strathclyde (including Cumberland and Westmorland), and the men of Alba (or Scots as the English sources call them). And it must be doubted that if the Viking force did so, no source would mention such a significant detail until John of Worcester.

The brief poem on the *Capture of the Five Boroughs*, following the poem on *Brunanburh* in the *Anglo-Saxon Chronicle* and on the same page in the *Parker Chronicle* manuscript (Cambridge, Corpus Christi College, MS 173, folio 27 recto), mentions the Humber as the boundary between the Danes in east Mercia and the Northumbrians of York. One might also ask why the poet of *Brunanburh* named the rather obscure *Dingesmere* as the route of escape for the Norsemen if he knew it was or was near the instantly and universally recognisable Humber.

Why does John come up with this idea? All the sources referring to the Humber entry derive from John, so the proposal that he had information from a lost early source has no independent corroboration. I suggest that John tended to think of invasions as coming via the Humber. He writes of *Brunanburh* (Livingston 2011: Darlington and McGurk 1995–1998: II, pp. 392–393): 'Hiberniensium multarumque insularum rex paganus Anlafus... ostium Humbre fluminis ualida cum classe ingreditur.' (Anlaf, the pagan king of the Irish and of many other islands … entered the mouth of the River Humber with a strong fleet.)

John writes of Harald Hardrada and Tostig's expedition, about 130 years later (Darlington and McGurk 1995–1998: II, pp. 602–603): 'Ad quem comes Tostius … sua cum classe uenit, et citato cursu ostium Humbre fluminis intrauerunt.' (Earl Tostig joined him with his fleet … and on a swift course they entered the mouth of the River Humber.) This reveals a kind of formulaic expression in John's account (Woolf 2007, p. 171 also points out that John recorded another Humber entry three years later); it perhaps also indicates that while John knew the name of the place, Brunanburh, he did not know its location.

It is suggested that the account of the battle in the *Annals of Ulster* (AU Mac Airt and Mac Niocaill, 1983) supports the idea of the Humber entry. The 937 annal reads (AU pp. 384–387):

> Bellum ingens lacrimabile atque horribile inter Saxones atque Norddmannos crudeliter gestum est, in quo plurima milia Nordmannorum que non numerata sunt, ceciderunt, sed rex cum paucis euassit.i. Amlaiph. Ex altera autem parte multitudo Saxonum cecidit. Adalstan autem, rex Saxonum, magna uictoria ditatus est.
>
> (A great, lamentable and horrible battle was cruelly fought between the Saxons and the Norsemen, in which several thousands of Norsemen, who are uncounted, fell, but their king, Amlaíb, escaped with a few followers. A large number of Saxons fell on the other side, but Athelstan, king of the Saxons, enjoyed a great victory.)

In the following year, 938, the *Annals* report Anlaf's return to Dublin (AU pp. 386–387): 'Amhlaiph m. Gothfrith i nAth Cliath iterum.' (Amlaíb son of Gothfrith in Áth Cliath [Dublin] again.) The delay in Anlaf's return is suggested to represent the length of time it took him to sail from the Humber to Dublin: 'Anlaf Guthfrith's son's arrival in Dublin "with a few" after the battle is recorded in the New Year of

---

* The first of these, 'The Brunanburh campaign: a reappraisal' (Halloran 2005), was discussed in Cavill (2008), but the second, 'The identity of *Etbrunnanwerc*' (Halloran 2010), was published too late to be given much attention in Michael Livingston's comprehensive volume, *The Battle of Brunanburh: A Casebook* (2011). As will be evident from the discussion below, no significant new evidence or argument relating to *Brunanburh* is brought forward by Halloran, and Livingston (2011) is likely to remain definitive.

938' (Wood 1980, p. 202), and again, 'If the fleet landed in the Humber, as is most likely, then it will have returned via Scotland, and the notice of its arrival at Dublin early in 938 would support this' (Wood 1980, p. 215, note 40). This is not quite accurate, since the *Annals* mention only Anlaf, not the fleet or the 'few' in 938 (see above), nor is it an inevitable inference.*

The 937 annal very closely follows the information given by the *Anglo-Saxon Chronicle* with the addition of circumstantial information about the awfulness of the battle, the numbers lost to the Norsemen, and the losses on the English side, none of which is beyond imaginative reconstruction. It is all written in Latin in a part of the *Annals* where Irish was becoming the norm (Dumville 1982). The Irish or hybrid Irish–Latin of the surrounding annals, including the reference to Anlaf returning to Dublin in 938, is typical of the *Annals* at this point of the text's development. This might suggest the annalist used a Latin source close to the *Anglo-Saxon Chronicle* for the *Brunanburh* entry and local knowledge for the 938 one.

The *Chronicle* poem indicates that Anlaf and a few followers escaped at a certain stage of the battle (Livingston 2011, pp. 40–41, lines 32b–36):

> Þær geflemed wearð
> Norðmanna bregu, nede gebeded
> to lides stefne litle weorode.
> Cread cnear on flot, cyning ut gewat
> on fealene flod. Feorh generede.
> (There was put to flight the Northmen's chief, driven by need to the ship's prow with a little
>   band. He shoved the ship to sea. The king disappeared on the dark flood. His own life he
>   saved.)

The poem does not say where Anlaf fled, and the language clearly indicates that one ship escaped by the use of the singular verbs and nouns. The *Annals*, having repeated the unnumbered dead of the Norsemen (Old English poem line 29, *unrim heriges* (countless men of the army); AU pp. 384–385, *que non numerata sunt* ([men] who are uncounted), likewise mentions Anlaf's escape with a small company (Old English poem line 34, *litle weorode* (with a little band); AU pp. 384–385, *rex cum paucis euassit* (their king escaped with a few followers) without saying where he fled. The Old English recounts a separate flight of the surviving Norsemen on *Dingesmere* back to Dublin, where the main verb and nouns and adjectives are plurals (Livingston 2011, pp. 42–43, lines 53–56):

> Gewitan him þa Norþmen nægledcnearrum,
> dreorig daraða laf, on Dingesmere
> ofer deop wæter Difelin secan
> eft Iraland, æwiscmode
> (Departed then the Northmen in their nailed ships, dreary survivors of the spears, on
>   Dingesmere, over deep water to seek Dublin, back to Ireland, ashamed in spirit.)

In the nature of the case, the poem is likely to be describing piecemeal flight rather than a disciplined and united withdrawal, and it is perfectly plausible that Anlaf might have escaped in a flight separate from that of his men, and as the poem implies, earlier than they did. The exigency of flight and the sense that his flight was an abandonment of men and responsibility might have caused Anlaf not to sail immediately to Dublin in the shame that the Old English poem so exultantly describes, but north perhaps, to stay with his allies until the dust settled. His return the following year might well have involved some negotiation to ensure his welcome at Dublin after the depleted forces had returned earlier and told

---

* Michael Livingston (personal communication, August 2011) notes that 'if such a delay is accepted for the aftermath of the battle, an even longer delay must also be accepted for the preparations leading up to the battle, and the historical record reveals that to be an impossibility. Anlaf Guthfrithsson took his army from a campaign in central Ireland to the field at *Brunanburh* in less than two months.'

their story. The sources do not tell us these things directly; but there is no particular reason, and no textual warrant, to suppose that the delay in Anlaf's arrival in Dublin recorded in the *Annals* had anything at all to do with a Humber entry of the fleet.

Michael Wood focuses on the putative Humber entry because he thinks 'we can be sure first of all that the object of the 937 invasion was Northumbria,' and that 'York was undoubtedly at the centre of these events, as it was in all the wars between the Norse of York and their allies on the one hand and the southern English on the other in the period 927–954' (1980, p. 201). Wood argues that the poem inserted by William of Malmesbury into his history shows that the Northumbrians submitted willingly to the invaders (1980, p. 201), and indeed that Northumbrians fought on their side in the battle. There can be no doubt that York was important, but this analysis fails to consider the composition of the invading forces and their joint objectives. While the Norse of Dublin were assuredly concerned with recreating the axis between Dublin and York, the submission of the Northumbrians (if true) would show their willingness for that too. Any ravaging in Northumbrian territory or Danish Mercia would likely be counterproductive, especially when rich pickings were available in west Mercia (Livingston 2011, p. 15, note 46). The force was not predominantly 'the Norse of York' but was a coalition of hitherto antipathetic groups, Scots and Cumbrians along with the Norse of Dublin and Northumbrians, who had all felt the sting of Athelstan's takeover of Northumbria in 927.

In the earlier years of Constantine's long reign, he and the Norse of Dublin fought each other at the battle of *Tinemore* or Corbridge in 918 when the Norse invaded the north. However, later events including Athelstan's northern expedition in 934 when he laid waste areas of Constantine's territory and forced his submission, made the formation of a coalition to withstand Athelstan's imperial pretensions imperative. Hatred of Athelstan was the motivating factor for the coalition, and 'Æthelstan's hegemony over the whole of mainland Britain … was threatened by an alliance' (Foot 2008, p. 133). The last thing the Scots and Cumbrians wanted was to substitute a Norse York- and Dublin-controlled Britain for an English-controlled Britain. The goal of the coalition was Winchester, not York. Ravaging English Mercia west of the Pennines would be part of the strategy and indeed that is what William of Malmesbury also tells us: *multum in Angliam processerat* (he [Anlaf] had advanced some distance into England) (Mynors, Thomson, and Winterbottom 1998–1999, I, pp. 206–207).

The tradition of the Humber entry cannot be traced earlier than John of Worcester and has been shown to reflect John's preconceptions about invading forces. There is little reason to suppose that the details recorded in the *Annals of Ulster* support the tradition of the Humber entry as they tally with those of the *Anglo-Saxon Chronicle* account and refer to the apparently independent movements of Anlaf and the fleeing Norse forces. A proper understanding of the objectives of the coalition forces would direct attention away from York and Northumbria as the targets of the invasion. There is very little to support the notion that the invading force came and left via the Humber.

Burnswark

Kevin Halloran's 2010 article, 'The identity of *Etbrunnanwerc*,' simply reasserts views that fail to engage with the evidence about the battle.[*] Early in his article, for example, he attempts to discredit the

---

[*] Some of these are referred to in the discussion of Burnswark below. My articles of 2007 and 2008 took issue with some of Halloran's ideas and arguments in his 2005 work and indicated where they were at variance with known fact or fuller evidence. Halloran's 2010 article selectively ignores both facts and evidence. He still maintains that his 'acceptance of the *burh* form [of *Brunanburh*] as genuine in no way detracts from the validity of … the suggestion that it was adopted for poetic purposes' (2010, p. 248, note 3), when it is a fact that a single alliterating stress in the first half an Old English poetic line is all that is required by the metre and, as I have shown, the battle was known as *Brunanburh* in prose texts where no alliteration is required (Cavill 2008, pp. 312–315). Halloran chooses 'correct' spellings of elements purely on the basis of what meanings he intends to attribute to them, so '*Wendune* rather than *Weondune* is the correct form and refers to the hill of Burnswark' (2010, p. 250) despite evidence to the contrary. He also decides that for the dative plural *nægledcnearrum* in line 53b of the Old English poem, 'the poet meant "in" rather than "to" the nailed ships' (p. 253, note 24) for no reason other than the need to align the sources to fit his proposed identification.

Bromborough case by quoting Campbell's comment, 'of course, the coincidence of *Brunan-*, a common, and *-burh*, a very common, place-name element, proves nothing relative to the site of the battle' (Halloran 2010, p. 249, note 6, quoting Campbell 1938, p. 59, note 4; cf. Wood 1980, p. 213, note 4). Certainly *-burh* is, as Campbell says, a very common element in place-names; but though I am aware of some place-names with the inflected element *brunan-* as the first element, it can hardly be called 'common.' Campbell was mistaken on this matter. But Halloran's logic is also at fault here. Even if *Brunan-* were a common element, and *-burh* very common, as Campbell supposed, then speaking from a merely statistical view, their co-occurrence in Bromborough would make that infinitely more probable as the location of the battle of *Brunanburh* than Burnswark, Brinsworth, Bromswold, or other candidates whose names contain neither of the elements. The fact is, however, that only Bromborough is reliably known to contain both elements, suggesting that naming it as a battle site is far less of a coincidence than has been thought.

Halloran begins by admitting that his earlier translation of the phrase 'apud Weondune quod alio nomine Etbrunnanwerc uel Brunnanbyrig appellatur' from Symeon of Durham as 'at *Weondune* which is otherwise named *Etbrunnanwerc* or called *Brunnanbyrig*' (2005, p. 145) was wrong. He now accepts that Symeon's phrase should be translated 'at *Weondune* which is called by another name *AetBrunnanwerc* or *Brunnabyrig*' (2010, p. 248 note 3). Symeon, in other words, makes clear by the use of the singular *alio nomine* (by another name) that he thought the names *Etbrunnanwerc* and *Brunnanbyrig* were for practical purposes synonymous, different versions of a single name.

Halloran nevertheless goes on to deny this result and to maintain that these variants refer not merely to different features but in fact represent the alterations of the name from *Brunnanbyrig* to *Etbrunnanwerc* (2010, p. 249). He proposes that the *burh* element refers to the fort on the summit at Burnswark and the *werc* element to the earthworks of the Roman camps on the slopes, undeterred by the fact that no evidence suggests that any part of Burnswark was historically referred to by *burh*, a term used in the area by English settlers, but not for this place (Barrow 1998, pp. 67–69).

Other commentators and I have suggested that a reasonable explanation of Symeon's phrase is that in some circumstances *burh* and *weorc* might both refer to fortified places or fortifications (Cavill 2008, p. 314; Woolf 2007, p. 171); certainly, they are used as synonymous variants frequently in Old English poetry. There is no doubt that the elements can historically refer to different aspects of the same place. These elements are not semantically inert, and where there is overlap, *burh* tends to refer to a fortified settlement and *weorc* to physical fortifications. Halloran sees the example I have given, culled from Smith's *Elements* (Newark Priory, originally *Aldebury*, Smith 1956, II, p. 254; Cavill 2008, p. 314, note 55) as 'questionable' and clearly doubts that the near-synonymy or partial overlap of these elements can be asserted, even though his entire argument ultimately depends upon it.

I have not conducted an exhaustive search, but the following may be offered as evidence for the co-existence and near-synonymy of names with *burh* and *(ge)weorc*. Simon Taylor's work, quoted by Halloran, gives the example of Southwark, also found on coin legends with *byrig*: 'there is some evidence to suggest that before the last type of Æthelræd II some coins reading SVÐB[yrig] were also struck at the Southwark mint, and that this was an alternate suffix for the place-name' (Smart 1981, p. 105; Taylor 1997, p. 17, note 31). Other examples include *Barnewerc* 13th century 'Beorn's fortification' in Burton Lazars, Leicestershire, which has the element *(ge)weorc* and Cox concludes 'this appears to represent the site of an early fortification, perh[aps] that from which Burton [*burh* + *tun*] took its name' (Cox 2002, p. 71). In East Bridgford, Nottinghamshire, near the Romano-British fort of *Margidunum*, known as *Aldwerch ca.* 1230 'old fortification' and Castle Hill is Burrow Fields 'from *burh*' (Gover, Mawer and Stenton 1940, p. 222).[*] Symeon, in common with these sources and their respective places, was most likely talking about the same place when he referred to 'Weondune quod alio nomine Etbrunnanwerc uel Brunnanbyrig appellatur.'

Halloran sees the 'problem' with my approach as that it 'relies too much on an analysis of forms that derive variously from copied, altered, difficult to read, and conflicting sources of uncertain provenance'

---

[*] A further example from a less documented survey, Buriton 'farm by the fortification' in Hampshire has a name War Down near Butser Hill 'perhaps containing OE *(ge)weorc* "(earth)-works"' (Coates 1989, pp. 44–45).

(Halloran 2010, pp. 252–253). This is the nature of the evidence and it has to be dealt with. Halloran prefers to make assumptions. Perhaps his biggest assumption is that the name Burnswark has anything to do with the battle of *Brunanburh*. I have shown that the earliest English sources and the overwhelming majority of spellings in all the sources clearly show that the first element of *Brunanburh* is a weak Old English substantive *bruna* or *brune* in the genitive singular *brunan* (Cavill 2008, pp. 303–309). Halloran's argument requires that we accept the spelling in just one manuscript of Gaimar (of four extant) from the early 14th century, *bruneswerce*, for the name (*Estoire des Engleis* line 3522; Short 2009, p. 192 (for text); pp. xix–xxii (for date). Bell (1960, p. 112) gives variant spellings (line 3518).

One 13th-century 'corrupt and contaminated' manuscript of Henry of Huntingdon (Greenway 1996, p. clxi) gave rise to a small number of late strong substantive spellings with *Brunes-* (Cavill 2008 and 2011, p. 349). The equation made by Halloran, '*Etbrunnanwerc* (*Bruneswerce*)' (2010, p. 251) is not made by any historical source and depends on grammatical and textual confusion on his part. The further implied equation, '*Etbrunnanwerc = Bruneswerce = Burnswark*' depends simply on the notion that the names look a bit similar (by the kind of logic, one supposes, that *suet* and *sweet* and *sweat* would equate). It also supposes that a nearly correct spelling of the place of the battle was first hit upon by the scribe of an Anglo-Norman poem in the 14th century and by no one else in the whole record.

Halloran directs attention to the *Burnswark* name and it is necessary to consider this name in some detail. I will first consider the new proposal broached in his article, that the first element of Burnswark might derive from a Celtic word *brïnn* (hill) in contradiction of his previous assertion that the *Burn-* of Burnswark cannot refer to a burn or burns because Neilson and others think it refers to *Bruna* (see below). Then I will re-examine my own suggestion for the etymology and bring forward some new evidence.

*Brunn

Halloran suggests a new etymology based on Alan James's generally excellent *Brittonic Language in the Old North* (2007 and continuing). Part of James's entry under *brïnn* (hill) reads:

> Burnswark Dmf (Hoddom) G Neilson in ScHistRev7 (1910), p39 n6 [**brïnn-**+ OE -*weorc* > "work"]; K Halloran pers. comm., and see ScHistRev84 (2005), pp133–48. If this was formed from a simplex **brïnn-**, it was in the "Pritenic" form **brun[n]-*.

This may be correct, and the gap of many centuries between the supposed formation of the name and the earliest extant spellings for Burnswark (given below from Neilson 1910 and Johnson-Ferguson 1936, with the earliest reference from 1541) may not argue against it. But any evidence to fill that gap of six centuries would add significantly to our understanding, and the absence of evidence makes circumspection appropriate. Halloran further quotes James as suggesting 'Burnswark could, then, have been a "Brit/Pict" simplex name *Brun[n]* taken up by Northumbrian English speakers to form *Brunes-weorc*' (2010, p. 251, note). There can be no serious argument about this because there is no evidence to take it beyond speculation.

I have no objection to the notion that 'there might be parallels for pre-English topographic names being used in OE place-name formation' (James, quoted by Halloran 2010, p. 251, note 16), or the evidence brought forward by Fox in 'The P-Celtic place-names of north-east England and south-east Scotland' (2007) that perhaps more names in the north derive from P-Celtic than have been hitherto recognised, as Halloran suggests. However, a further factor bears on the plausibility of the suggestion that Burnswark might derive from *brïnn*. Another part of James's entry relates to the etymology of the word: 'the root **bhreu-* is associated with "swelling" in various senses, and the close affinity between this word and that for "breast" (see **bronn**) may indicate the characteristic shape of a **brïnn**, "hill"' (see also VEPN under **brunnjo-*, 'ModW *bron* "breast, hill" is a related term … which is also common in Welsh p.ns.'). Burnswark is an extraordinary looming plateau with very steep sides and a flattish top (see the cross-section in Roy 1793 and Figure 6.1). Given that all hills are some types of swellings or protuberances, a less breast-like hill than Burnswark would be hard to imagine.

FIGURE 6.1   Burnswark, Dumfriesshire, from the south—not 'a breast-shaped hill.'

If we were to suppose that **brunn* could be the etymon of the first element of Burnswark, the fact remains that nothing whatever in that proposed etymology identifies Burnswark with Brunanburh. James points out that **brunn* may also be the etymon of other names in the north. As a term for a topographical feature, it may also refer to many a hill now with another name. If we suppose that the phrase in one manuscript of the *Annales Cambriae*, 'Bellum Brune,'* actually refers to a **brunn* generically, it could refer to any appropriate hill. Certainly if it were to refer to a named place, a specific hill, there is no particular reason to suppose it to refer to Burnswark rather than, say, Bryn in Lancashire, The Brinns in Westmorland, or any of the other possible examples of *brïnn* or **brunn* cited in James's work.† If this etymology were accepted, it would be a peculiarly non-referential annal in terms of the location of the battle, amounting to no more than 'battle of the [or a] hill' or 'battle of Hill.'

Do the *Annales Cambriae* in fact refer to a **brunn*? The actual evidence can be read in different ways, although it might be noted that most references to battles in the *Annales* mention named places rather than generic topographical features.‡ Scholars of P-Celtic tend to think that the entry does not refer to a **brunn*. Andrew Breeze suggests that it was English: '*Brun(e)* was a place-name known to the Welsh,' and again 'the English name *Bruna* or *Brune* was presumably understood by the Welsh as meaning "stream"' (1999, p. 481). John Bollard and Marged Haycock read the **kattybrunawc* of the late 10th-century Welsh *Glaswawt Taliessin* as 'the battle for the settlement in Brun's region,' the *Cad Dybrunawc* of the late 12th-century *Canu y Dewi* as 'the battle of Brunanburh(?)', and the '*Ac y bu ryfel Brun*' in the late 13th-century *Brut y Tywysogion* as 'and there was the battle of Brun' (Livingston 2011, pp. 48–49, 66–67, and 88–89, respectively). The consensus here is that this *Brun* and *-brun-* were probably references to the English element in *Brunanburh* and not a meaningful Celtic element where one would expect such an element to be recognised.

---

* The actual manuscript of the *Annales Cambriae* that contains the reference to *Brune* (London, British Library, MS Harley 3859) is early 12th-century. The *Annales Cambriae* text is 'interpolated … into a copy of … *Historia Brittonum*'; it took material from an Irish chronicle of the first half of the 10th century. The Harley manuscript is the only text to record 'Bellum Brune' whereas all the other battles (except possibly *Conani*, 881; see next note) in the period of vernacular annals, 682–954, figure in at least two. This suggests that *Brune* might have been a battle recorded in an independent tradition unknown, or of no interest to the scribes of the other manuscripts.

† It might be noted that Ekwall 1922, p. 100 accepts 'W. bryn' as the etymon of Bryn, but Smith 1967, II, p. 173 is not inclined to think that The Brinns derives from *brïnn*.

‡ In the section of the texts 682–954 edited by Dumville (2002), there are references to battles in 722 *bellum Hehil*, *gueith Gart Mailauc* and *cat Pencon*; in 728 to *Bellum Montis Carno*; in 750 to *gueith Mocetauc*; in 760 to *Bellum … gueith Hirford*; in 796 to *bellum Rudglann*; in 844 to *Gueith Cetill*; in 848 to *Gueit Finnant*; in 870 870 to *Cat Brinonnen*; in 874 to *Gueith Bannguolou*; in 877 to *Gueith diu Sul in Mon*; in 881 to *Gueit Conguoy*; in 906 to *Gueith Dinmeir*; in 921 to *Gueith Dinas Neguid*; in 937 to *Bellum Brune*; and in 951 to *bellum Carno*. Most of these places have been identified.

The point, then, is first that a supposed **brunn* in Burnswark does not in any sense enforce a link between 'Bellum Brune' and that place; and second, that good arguments have been made for the *Brune*, *Brun*, *-brun-* to refer to the English element. Halloran's speculations make for a false syllogism which runs something like this: 'the etymon of the first element of Burnswark might be **brunn* (a hill); Celtic sources might refer to a **brunn* (a hill) as the scene of the battle; therefore Burnswark was the site of the battle.' Neither of the propositions is demonstrably true and the conclusion manifestly does not follow. As far as determining the site of the battle is concerned, the proposed new etymology of Burnswark is a red herring.

Burn

Halloran pours scorn on my observation that Burnswark might actually be derived from Scots or Middle English *burn* (a stream) and *wark* (a fortification), calling these possibilities 'unfeasibly late' and asserting that I give 'no detailed etymology' for my suggestion (2010, p. 252). He prefers to ask 'what were the original names for the hill and the hill-fort?' (2010, p. 252). Sadly, he does not tell us the answer to his question, unless it is something to do with the posited **brunn* element discussed above.

I have no doubt that the hill-fort was called something else earlier in its existence, but what that name was and whether the name had any relation to the present one is anybody's guess. We can only wait for the evidence to be found. For the present, we must deal with the evidence we have.

Halloran refers to the opinions of Neilson (1910), Johnston (1934) and Mills (2003) about the meaning of the name[*] to dismiss the notion that the early spellings of Burnswark offered by Neilson, namely *Burnyswarke* 1542, *Burniswork* 1608, *Burneswark* 1623, and *Burnswark* 1661 might refer to a burn or burns. Johnson-Ferguson adds two further early spellings, *Burniswerkhill* 1541 and *Burniswarkleyis* 1625 (1930, pp. 54–55). Halloran cites one additional spelling of *Brunswork* from Sir John Clerk in 1730 as if this negates or cancels out the multiple earlier *Burn-* spellings (2010, p. 252).

Halloran appears to think that the 'presence of "burn" names for a few farmsteads in the locality' (and, as I noted, for the streams in the vicinity, of which there are many) somehow makes it especially implausible that there is a reference to *burn* in the name Burnswark, as reflected in the earliest spellings we have. Likewise he argues that the complete absence of *dun* names nearby and the complete absence of characteristic *dun* features make Burnswark a plausible *dun* (2010, p. 252; see further on **brunn* above). This is, to say the least, a counterintuitive use of topographical evidence.

In terms of the detailed interpretation of the spellings, the *Dictionary of the Scottish Language* (DSL) gives plenty of evidence for *burn* '1. A brook or stream,' and for *wark* '9.a. The (action of or activity concerned with) building, repairing, etc. (*of* an edifice, etc.).' The northern Middle English and Scots plurals and genitive singulars of nouns are typically *-ys*, *-is*, and *-es*. Both dictionary entries specifically mention the use of these words in place-names, so my proposed etymology 'fortification of the burn, fortification (in the area) of the burns' clearly reflects this particular strand of historical, geographical and linguistic evidence. This interpretation is entirely plausible if the name Burnswark represents a late Middle English formation.

Birrenswark

An alternative explanation for the name derives from the persistent but not yet fully documented form *Birrenswark*. This is listed by Johnston (1934) and Johnson-Ferguson (1936) as the headform of the name; is depicted as 'Plan and sections of Birrenswork-hill' as early as 1793 in Plate XVI of William Roy's *Military Antiquities of the Romans in North Britain*; and is mentioned by Christison (1899) and

---

[*] The only one of these interpreters able to revise his view on the meaning of Burnswark in the light of the evidence, David Mills, has done so: 'I shall certainly want to change the entry for Burnswark in my Dictionary (this should be possible when they next do a reprint). I'm inclined to be a bit cautious about the first element, so will probably have something like: ME wark (OE weorc) "fortification" (here referring to a Roman camp), first element uncertain, possibly ME burn (OE burna) "spring"' (personal communication, January 2011).

the *Scottish National Dictionary*. In the dictionary, *birren* is glossed 'a camp,' and it is noted, 'Occurs in the proper name *Birrenswark* in D[u]mf[ries]sh[ire].' The element, from Old English *burgæsn*, is 'a northern term, especially common in We[stmorland]' (VEPN), but also known in Dumfriesshire. The *Scottish National Dictionary* records it in use in 1834 in Dumfriesshire with a quotation reading, 'small entrenched camps or *Birrens*, as they are called.' The explanatory gloss in the quotation indicates that the writer was not certain that his 19th-century readers would understand the term. The word is found in Dumfriesshire place-names, e.g., a Birrens Hill at NY 2481 and a Birrens Sike at NY 3992.

Perhaps more interesting is the fact that the Roman fort of Blatobulgium, just outside Ecclefechan at NY 218752, is called Birrens, at least from 1793 in Roy's work mentioned above and continuing into the present (Robertson 1975). The interest of this name is that Burnswark or Birrenswark was clearly an outpost of the garrison at Birrens–Blatobulgium, and it is likely that here we have a good illustration of Taylor's model of name development: 'There existed originally a core simplex name.... This core simplex name referred in general terms to an estate or area of land perceived as some kind of entity. To this simplex could then be added elements defining the particular aspect of that entity which the speaker wished to single out' (1997, p. 11). Birrens, the English name of the station of Blatobulgium, is the simplex form that may well have given rise to the compound Birrenswark, 'the fortification or earthworks on the Birrens estate, the land commanded by Birrens,' with the defining generic *wark* (fortification, earthworks, camp).

Birrens and the derived Birrenswark descend from Old English *burgæsn* (burial, cairn) and here we may have the evidence for 'the original names for the hill and the hill-fort' that Halloran guesses at (2010, p. 252). This takes the evidence back some way at least.

It is possible, furthermore, that the *birrens* element of Birrenswark was assimilated to the more familiar *burns-* of the present Burnswark and the early spellings. This process can be documented for Westmorland names deriving from *burgæsn*, namely Swathburn (*Swarthburchanes* 1295), Griseburn (*Griseburghanes ca.* 1216), and Mossburn (*Mosburhannes* 1291), all documented in VEPN under *burgæsn*. This may well explain the early *burn(e, y, i)s-* forms of the present Burnswark name discussed above and makes a plausible link with the main variant of the name.

The case for Burnswark as the site of the battle of *Brunanburh* has not been advanced by the addition of more speculation to the extremely speculative suggestions already proposed by Halloran. Michael Livingston (2011, p. 19, note 59) observed, for example, that his interpretation of the military and tactical aspects of the campaign that form a major plank in Halloran's argument (2005, pp. 138–140; 2010, p. 248, note 4), is nowhere as secure as he suggests. More thorough examination of the onomastic evidence relating to Burnswark makes any connection with *Brunanburh* vanishingly remote.

Dingesmere

If we do not trust the explicit link between *Brunanburh* and Bromborough, one other name in the Old English poem may help locate the battle. This is *Dingesmere*, the place from which the Dublin Vikings fled by ship after the rout. Campbell rightly recognised it as a key piece of information for identifying the battle site. In a recently-published article colleagues and I argue that *mere* is the place-name element meaning 'wetland, land which is subject to flooding' and that the *ding* is an English reflex of Old Norse *þing* (local council), well exampled in English names and more widely in names such as Dingwall in Inverness (Cavill, Harding and Jesch 2004; Cavill 2007).[*]

We suggested that the name made reference to the *þing* of Thingwall, the meeting place of the Viking settlers in the Wirral. *Dingesmere* would then be the wetland, the muddy estuary regularly flooded and drained by the Dee, and overlooked by the Thing at Thingwall (Figure 6.2). If then, as the poem says, the survivors of the Norsemen went to or in their ships 'on *Dingesmere*, over the deep water back to Dublin,' an area somewhere along the Dee coast for this escape is plausible. The politically independent Norse enclave in the north and west of the Wirral in the 10th century, and woodland in the north of the

---

[*] Following an initial suggestion by Steve Harding that *Dingesmere* may possibly derive from 'Thing's mere.'

FIGURE 6.2 *Dingesmere?* Heswall Point on the Dee estuary at low tide. In the distance is the coast of north Wales, and between there and the bank dividing the picture is the Dee. At the highest tides, the water reaches beyond the vegetation at the bottom of the picture. The small water outlet flows from left to right and meets the estuary proper about 500 m right of the area at the border of the picture. The coastline has changed since the Middle Ages, but the usefulness of the area for beaching craft is visible even today. (Photo courtesy of Steve Harding.)

peninsula, might have been thought to be more hospitable to Vikings fleeing from Bromborough than the English garrison at Chester.[*]

One of the difficulties in assessing the welter of suggestions about the location is the fact that arguments are made on the basis of selected and often out-of-date opinions rather than evidence. The careful argument about *Dingesmere* already mentioned has been dismissed because, in desperation, Joseph Bosworth glossed *On dinnes mere* as 'on a stormy sea' and *On dynges mere* as 'on the sea of noise' (Bosworth and Toller 1898, pp. 205, 221 under headwords *dinne* and *dynge*). It should be noted that the headwords and definitions relate only to the name occurring in the *Brunanburh* poem, and there are no other quotations to exemplify the words.

None of those opposing the 'wetland of the thing' interpretation answered these real objections to Bosworth's glosses, that Toller in 1921 could see when he excised the whole *dynge* entry and suggested that *dinnes* 'seems to point to a proper name' (Bosworth and Toller *Supplement* 1921, p. 162). The least technical of these objections is the fact that neither the proposed **dinne* (storm or tempest) nor **dynge* (noise, dashing, or storm) (see Halloran 2010, p. 253, note 24) exists in Old English according to the *Dictionary of Old English*, the most comprehensive linguistic collection and analysis of Old English. They are in fact ghost words.

The variant spellings of the word in the Old English poem manuscripts are as follows:

ASC A, Cambridge, Corpus Christ College, MS 173, *dinges mere*

ASC B, London, British Library Cotton Tiberius A. vi, *dyngesmere*

ASC C, London, British Library Cotton Tiberius B. i, *dinges mere*

ASC D, London, British Library Cotton Tiberius B. iv, *dynigesmere*

[ASC O, London, British Library Cotton Otho B. xi (lost), *dinnesmere*].

The ASC A and C manuscripts agree on the spelling *dinges-*; B's *dyngesmere* is a spelling variant (*y* for *i* is common in late West Saxon); the D reading with *dyniges-* might have been thought to mean 'mere of

---

[*] Sources such as *Egilssaga* must be used with caution, but it is worth noting that Keith Kelly discerns an echo of Chester in the saga's reference to Athelstan's troops occupying a fortress south of the battle site (Livingston 2011, p. 210).

the wild parsnip' (see DOE under *dynige*, although there is only one example of the word, among a list of plants). The last, the O spelling *dinnesmere*, is only extant from a 16th-century copy of a copy of A and is of no independent evidential value (Campbell 1938, pp. 1, 115, 133–144). It must be regarded as a misreading, transcription error, or 'an alteration by the scribe' (Campbell 1938, p. 115). Thus while the O scribe or his 16th-century counterpart might 'have meant *dinnes* as gen. s. of *dyne* "noise"' (Campbell 1938, p. 115), and indeed Campbell might have 'considered it a reasonable explanation for the *Dinnes*-form' (Halloran 2010, p. 253, note 24), that can have no 'value in determining ... the text ... of the poem' (Campbell 1938, p. 1)—or its meaning.

The spelling with *-g-* is persistent, and that rules out a form of the verb *dynian* (to resound) and the noun *dyne* or *gedyne* (din, loud noise). The feminine noun *dyncge* (dung, manure) has been suggested in some quarters, but this only occurs in glosses, apart from a word appearing in a law text as *ðingan* (DOE). In neither the weak (Laws example) nor the strong (glosses) forms would the genitive singular inflection be *-(e)s*. Some early glosses (DOE Corpus) spell *þing-* forms *ding-*, for example *quoquemodo – aengi dinga* (in any way), *aduocatus – dingere, dingare* (counsellor). An English etymon for *dinges*—is thus hard to find.

The Norse language yields no better source for the *dinges-* spellings, The verb *dynja* (to gush, shower, pour) gives no *-g-*; *dengja* (to whet) is rare and not spelt with *-i-* or *-y-*; and the noun *dyngja* (lady's bower) (Cleasby, Vigfusson, and Craigie 1957, pp. 111, 99, and 111, respectively) may with some ingenuity approach the Old English spellings, but the word would not make much sense and certainly would not mean 'din'. The idea that *Dingesmere* means 'sea of noise' or something similar is thus founded on no valid evidence.

A more technical objection, but important nevertheless, is the fact that, as I have shown, the simplex *mere* and *mere* as the first element of compounds unproblematically refer to the sea or an expanse of water in verse, including Grendel's mere in *Beowulf*. But as the second element of a compound in verse or in place-names, *mere* does not denote 'sea'; in poetry its core meaning is 'pool.' In place-names, its core meaning is 'pool, wetland' (Cavill 2007, pp. 35–38).

We have many examples of place-names with qualifiers in the genitive and the generic *mere*, like *dingesmere*, but no non-toponymic poetic compounds with such a configuration of 'specific in the genitive + *mere*', thus, in the configuration in which it appears in the poem, the evidence suggests first that *dingesmere* is a place-name and second that it refers to a (coastal) wetland, not to the sea. The appropriateness of that description to the estuary of the Dee at Heswall and overlooked by Thingwall has already been outlined.

Painful though it may be, we must reckon with a wide range of linguistic, stylistic, and topographical evidence to arrive at conclusions about words like *dingesmere*. The analysis of the evidence that I have presented shows that the word refers to neither noise nor sea. The interpretation of *dingesmere* as 'wetland of the *þing*' (Cavill, Harding and Jesch 2004) may not yet be unassailable, but it has not been effectively challenged by appeals to outdated guesswork.

Conclusion

This book brings together a wealth of evidence relating to the North Sea coasts and the Chester area, showing their significant roles in war and trading, even if they did not feature very prominently in the written sources of the first millennium. I have tried to give a brief account of where the debate about *Brunanburh* stands at present, and in so doing have had to interact in detail with opinions that place the battle elsewhere than the Wirral. I have not been able to include and discuss all the material that is added daily on the Internet. New interpretations are constantly being proposed, but ultimately they all float or founder on the evidence.

I believe that that evidence decisively points to the battle of *Brunanburh* having been fought on the Wirral, and that careful scholarship locates this particular encounter between the English and the Vikings most plausibly at Bromborough. We have come far in our search for the battle: to Campbell's *nescio* we can now say, with some confidence, *puto scio*.

ACKNOWLEDGMENTS

I would like to thank the organisers of the Chester conference for the stimulating programme and smooth arrangements. Michael Livingston read this chapter and suggested some changes and additions, to its great benefit, and I thank him; any errors are my own.

REFERENCES

AU—*Annals of Ulster*, see Mac Airt and Mac Niocaill.

Barrow, G.W.S. (1998). The uses of place-names and Scottish history: pointers and pitfalls. In *The Uses of Place-Names*, S. Taylor, Ed. Edinburgh, Scottish Cultural Press, pp. 54–74.

Bell, A. (1960). *L'Estoire des Engleis by Geffrei Gaimar.* Anglo-Norman Texts XIV–XVI. Oxford, Blackwell.

Bollard, J.K. and Haycock, M. (2011). The Welsh sources pertaining to the battle. In *The Battle of Brunanburh: A Casebook*, M. Livingston, Ed. Exeter, University of Exeter Press, pp. 245–268.

Bosworth, J. and Toller, T.N., Eds. (1898; reprinted 1973). *An Anglo-Saxon Dictionary*. London, Oxford University Press.

Bosworth, J. and Toller, T.N., Eds. (1921; reprinted 1980). *An Anglo-Saxon Dictionary: Supplement* with revised and enlarged addenda by A. Campbell. London, Oxford University Press.

Breeze, A. (1999). The battle of Brunanburh and Welsh tradition. *Neophilologus* 83, 479–482.

Campbell, A. (1938). *The Battle of Brunanburh*. London, William Heinemann.

Cavill, P. (2007). Coming back to *Dingesmere*. In *Language Contact in the Place-Names of Britain and Ireland*, P. Cavill and G. Broderick, Eds. Nottingham, English Place-Name Society, pp. 27–41.

Cavill, P. (2008). The site of the battle of *Brunanburh*: Manuscripts and maps, grammar and geography. In *A Commodity of Good Names: Essays in Honour of Margaret Gelling*, O.J. Padel and D.N. Parsons, Eds. Stamford, Shaun Tyas, pp. 303–319.

Cavill, P. (2011). The place-name debate. In *The Battle of Brunanburh: A Casebook*, M. Livingston, Ed. Exeter, University of Exeter Press, pp. 327–349.

Cavill, P., Harding S.E., and Jesch J. (2004). Revisiting Dingesmere. *Journal of the English Place-Name Society* 36, 25–38.

Christison, D. (1899). Account of the excavation of the camps and earthworks at Birrenswark Hill, in Annandale. *Proceedings of Society of Antiquaries of Scotland*, 34, 198–218.

Cleasby, R., Vigfusson, G., and Craigie, W. (1957). *An Icelandic–English Dictionary*, 2nd ed. with Supplement. Oxford, Clarendon Press.

Coates, R. (1989). *The Place-Names of Hampshire*. London, Batsford.

Cox, B. (2002). *The Place-Names of Leicestershire, Part 2*. Nottingham, English Place-Name Society.

Darlington, R.R. and McGurk P., Eds. (1995–1998). *The Chronicle of John of Worcester,* 3 vols. Oxford Medieval Texts. Oxford, Clarendon Press.

Dictionary of Old English and *Dictionary of Old English Corpus* http://tapor.library.utoronto.ca/doe/dict/index.html, referred to by headword.

Dictionary of the Scottish Language (incorporating *Dictionary of the Older Scottish Tongue* and *The Scottish National Dictionary*) http://www.dsl.ac.uk/index.html

Dodgson, J. McN. (1957). The Background of Brunanburh. *Saga-Book of the Viking Society* 14, 303–316. Reprinted in Dodgson, *The Place-Names of Cheshire*, 5.2. pp. 249–261, and cited from this edition.

Dodgson, J. McN. (1970–1997). *The Place-Names of Cheshire*, 5 parts in 7 vols. Nottingham, English Place-Name Society and Cambridge, Cambridge University Press.

Downham, C. (2008). *Viking Kings of Britain and Ireland: The Dynasty of Ivarr to AD 1014*. Edinburgh, Dunedin Academic Press.

Dumville, D.N. (1982). Latin and Irish in the Annals of Ulster, A.D. 431–1050. In *Ireland in Early Mediaeval Europe: Studies in Memory of Kathleen Hughes,* D. Whitelock et al., Eds. Cambridge, Cambridge University Press, pp. 320–341.

Dumville, D.N. (2002). *Annales Cambriae, A.D. 682–954: Texts A–C in Parallel*. Cambridge, Department of Anglo-Saxon, Norse and Celtic.

Ekwall, E. (1922). *The Place-Names of Lancashire*. Manchester, Chetham Society.

Foot, S. (2008). Where English becomes British: rethinking contexts for *Brunanburh*. In *Myth, Rulership, Church and Charters: Essays in Honour of Nicholas Brooks*, J. Barrow and A. Wareham, Eds. Aldershot, Ashgate, pp. 127–144.

Foot, S. (2011). *Æthelstan the First King of England*. New Haven, Yale University Press.

Fox, B. (2007). The P-Celtic place-names of north-east England and south-east Scotland. *The Heroic Age* 10. http://www.mun.ca/mst/heroicage/issues/10/fox.html

Gover, J.E.B, Mawer, A., and Stenton, F.M. (1940). *The Place-Names of Nottinghamshire*. Cambridge, Cambridge University Press.

Greenway, D., Ed. (1996). *Henry Archdeacon of Huntingdon: Historia Anglorum, The History of the English People*. Oxford Medieval Texts, Oxford, Clarendon Press.

Halloran, K. (2005). The Brunanburh campaign: a reappraisal. *Scottish Historical Review* 84, 133–148.

Halloran, K. (2010). The identity of *Etbrunnanwerc*. *Scottish Historical Review* 89, 248–253.

James, A. *Brittonic Language in the Old North*. http://www.spns.org.uk/bliton/Aindex.html

Johnson-Ferguson, E. (1936). *Place Names of Dumfriesshire*. Dumfries, Courier Press.

Johnston, J.B. (1934). *Place-Names of Scotland*. 3rd ed., reprinted 1970. London, S.R. Publishers.

Livingston, M., Ed. (2011). *The Battle of Brunanburh: A Casebook*. Exeter, University of Exeter Press.

Mac Airt, S. and Mac Niocaill G., Eds. (1983). *The Annals of Ulster*. Dublin, Institute for Advanced Studies.

Mills, A.D. (2003). *A Dictionary of British Place-Names*. Oxford, Oxford University Press.

Mynors, R.A.B., Thomson, R.M., and Winterbottom, M. Eds. (1998–1999). *William of Malmesbury: Gesta Regum Anglorum*, 2 vols. Oxford Medieval Texts, Oxford, Clarendon Press.

Neilson, G. (1909). Brunanburh and Burnswark. *Scottish Historical Review* 7, 37–39.

Page, R.I. (1982). A Tale of Two Cities. *Peritia* 1, 335–351.

Parsons, D. and Styles, T. with Hough C. (1997 and continuing). *The Vocabulary of English Place-Names*. Nottingham. English Place-Name Society.

Robertson, A.S. (1975). *Birrens (Blatobulgium)*. Edinburgh, Constable.

Rollason, D. (2000). *Symeon of Durham. Libellus de exordio atque procursu istius, hoc ecclesie: Tract on the Origins and Progress of this the Church of Durham*. Oxford Medieval Texts. Oxford, Clarendon Press.

Roy, W. (1793). *Military Antiquities of the Romans in North Britain*. London, Society of Antiquaries.

Short, I. (2009). *Geffrei Gaimar, Estoire des Engleis/History of the English*. Oxford, Oxford University Press.

Smart, V. (1981). *Sylloge of Coins of the British Isles 28: Cumulative Index of Volumes 1–20*. London, British Academy.

Smith, A.H. (1956). *English Place-Name Elements*, 2 vols. Cambridge, Cambridge University Press.

Smith, A.H. (1967). *The Place-Names of Westmorland*, 2 vols. Cambridge, Cambridge University Press.

Taylor, S. (1997). Generic-element variation, with special reference to eastern Scotland. *Nomina* 20, 5–22.

Taylor, S., Ed. (1998). *The Uses of Place-Names*. Edinburgh, Scottish Cultural Press.

VEPN = *The Vocabulary of English Place-Names*, see Parsons and Styles, referred to by headword.

Wood, M. (1980). Brunanburh revisited. *Saga-Book of the Viking Society* 20, 200–217.

Wood, M. (1999). *In Search of England: Journeys into the English Past*. London, Viking.

Woolf, A. (2007). *From Pictland to Alba, 789–1070*, The New Edinburgh History of Scotland, Vol. 2. Edinburgh, Edinburgh University Press.

7

Viking Age Rural Settlement in Lowland North-West England: Identifying the Invisible?

Robert A. Philpott

CONTENTS

ABSTRACT Recent surveys have highlighted the lack of research into the early medieval rural archaeology of north-west England. From the end of the Roman period until well after the Norman Conquest, the very sparse material culture and lack of recognisable cropmark signatures make it very difficult to locate rural settlements and, once found, to establish a chronology. As a result, most early medieval activity has been found accidentally when excavating sites of earlier or later periods.

Two sites of early medieval date, including Viking Age phases, excavated in Wirral since the late 1980s, illustrate the challenges facing those studying the early medieval archaeology of the region. They show that, despite the scarcity of finds, distinctive building types combined with detailed stratigraphical analysis and radiocarbon dates can offer the potential to recognise and characterise elusive early medieval occupation.

Identification of sites in the wider landscape requires close attention to the subtle clues that do exist including place-names, field names, and primary field enclosures. Accurate plotting of find spots of metal and other objects may over time build up into significant concentrations of finds. In modern developments, extensive stripping of areas (the main method of finding such sites), indicators such as undated features, dark soils, and concentrations of non-diagnostic fired clay may be the only indicators of early medieval settlement.

Introduction

The problem of identifying Viking Age rural sites afflicts rural settlements of the entire early medieval period over many parts of western Britain (Thomas 2012, pp. 43–44). The lowland north-west England (modern southern Lancashire, Cheshire, Merseyside, and Greater Manchester) exemplifies the 'problem' and research challenge. Rural sites that were occupied within a span of 800 years from the end of Roman administration around AD 400 to the 13th century are notoriously difficult to locate. The virtual absence of pottery between the Roman occupation and the High Middle Ages renders site identification extremely difficult (Higham 2004a, p. 23).

Other regions such as parts of the West Midlands and Wales share a similar problem in locating early medieval sites. One of the nearest Anglo-Saxon rural settlements to Wirral in the Trent Valley, the settlement of Catholme, Staffordshire spanning the early 6th to the 10th century yielded no identifiable pottery on the surface and was discovered only through recognition of three apparent *grubenhäuser* (sunken-floored buildings) in aerial reconnaissance, two of which on excavation proved to be peri-glacial features (Losco-Bradley and Wheeler 1984). In addition to a virtually aceramic material culture, there is a general dearth of other diagnostic artefacts that, when combined with the relatively ephemeral building forms and settlement remains in physical terms, renders site location highly problematic.

These difficulties have inhibited research in north-west England, as recent national surveys have recognised. Stuart Wrathmell (2012, p. 258) observed in a review of medieval settlement in northern England that, with the notable exception of Tatton, "it is hard to refute the claim that north-west England as a whole has been largely ignored by those concerned with medieval settlement research."

The early medieval period is least well-served. Gabor Thomas (2012, p. 59) explicitly identified north-west England as one of the most obvious gaps in the uneven distribution of the archaeological resource for early medieval settlement, noting that 'one of the imperatives of future research must be to generate more high-quality data for the black holes.' Helena Hamerow (2012, p. 32) lamented that 'regrettably few "ordinary" farmsteads of the 8th to 10th centuries in north-west England have been identified and still fewer excavated.'

Against this undistinguished yet challenging background, this chapter will examine some of the problems of identifying Viking Age settlements and those of early medieval date more generally in the lowland north-west of England. It will present some of the archaeological evidence from sites where early medieval settlements have been located, look at problems of chronology and site recognition, and draw out some of the implications that may assist in site location in the future.

Known Sites

A small but significant number of sites where early medieval structures or activity have been identified exist in the region (Table 7.1; Figure 7.1). However, as the table illustrates, the early medieval components often form relatively minor elements of reports that focus on sites of other periods. In most cases, the crucial dating evidence comes from radiocarbon determinations.

The sites are widely dispersed geographically and diverse chronologically but some common themes are considered below. One characteristic of most of the early medieval phases including the two Wirral examples at Irby and Moreton is that they were accidentally discovered in searching for or investigating sites of other periods (Higham 2004a, p. 23). While this provides a means of finding some rural early medieval sites, it obviously fails to locate those that for various reasons do not occupy the same locations as those of more archaeologically visible or materially more productive periods.

Glimpses of the Viking period in excavations are otherwise rare, with isolated radiocarbon dates or individual finds mainly providing the small stock of existing knowledge. Two case studies of Wirral sites where Viking Age evidence has been recovered exemplify some of the problems of finding sites, but they also provide a basis for understanding Viking Age rural settlement in the region and investigating the techniques of site location.

Irby

The site at Irby, Wirral (Figure 7.2), provides us with an example of long-lived occupation that includes the early medieval period (Philpott and Adams 2010). In 1987 Roman sherds found in building work led to the discovery of a Romano-British enclosure ditch. Subsequent extensive excavations from 1992–1996 funded by National Museums and Galleries on Merseyside, English Heritage, and the University of Liverpool, revealed a multi-period rural settlement occupied during the Bronze Age and Iron Age,

TABLE 7.1

Some Excavated Early Medieval Rural Sites in Lowland North-West England

Site	Dating	How Discovered	References
Tatton, Cheshire	Late Roman/early sub-Roman radiocarbon dates	In excavation of deserted medieval village	Higham 2004b, p. 308
Helsby, Cheshire	C14 dates from build-up against prehistoric rampart AD 400–530	C14 dates on Iron Age hillfort	Garner 2012, p. 65
Hilary Breck, Wallasey, Wirral	5th to 6th century	Targeted for potential early medieval activity, confirmed by C14 dates	Adams 2012
Birch Heath, Tarporley, Cheshire	C14 dates from foundation gully of structure AD 590–720	C14 dates on Romano-British farmstead	Fairburn 2002, pp. 71–73
Court Farm, Halewood, Knowsley	C14 ca, AD 680–980 (Beta-108098 1210 ± 60BP) from ditch cutting Roman deposits	C14 dates on Romano-British hamlet excavation	Adams and Philpott (in preparation)
Timperley Old Hall near Altrincham, Trafford	Associated with fence-line and hearth, the latter producing C14 date of 8th to 9th century AD	C14 dates on features at medieval and post-medieval hall	Nevell 1997
Brook House Farm, Halewood, Knowsley	C14 date of ca. AD 1000–1230 (930 ± 50BP; Beta-117713) from layer in infilled inner ditch	C14 dates excavation within Iron Age and Romano-British enclosure; stratigraphy	Cowell and Philpott 2000
Tatton, Cheshire	'Late Saxon or Viking-age' associated with Chester ware sherd	In excavation of deserted medieval village	Higham 2004b, p. 310
Manchester	4 pits (stratigraphic position)	In excavation of Roman fort	Bryant et al. 1986, pp. 54–56, Figure 4.27
Irby, Wirral	See Irby section in this chapter	In excavation of Romano-British and multi-period site	Philpott and Adams 2010
Moreton, Wirral	See Moreton section in this chapter	In excavation for medieval chapel	Philpott (in preparation)
Cow Grange Worth, Ellesmere Port	Two Chester ware sherds	In excavations at moated monastic grange of Stanlow Abbey	Brotherton-Radcliffe 1975, p. 78, Figure 5.1
Castle Ditch, Eddisbury, Cheshire	Archaeomagnetic date of AD 750–1000 on clay oven	Research excavation on Iron Age hillfort	Garner 2012, p. 66
Kelsborrow Castle, Cheshire	Ditch fill dated AD 1020–1160; pit dated AD 990-1120; another pit dated AD 690–890	Research excavation on Iron Age enclosed settlement	Garner 2012, p. 67

followed by a Romano-British enclosed farmstead that may have remained in use into the post-Roman period, and then saw early medieval and later medieval occupation.

It proved a difficult site to excavate, with 46 separate trenches spread across 8 gardens. The great majority of the structural evidence is of Bronze Age, Romano-British, or later medieval date. The long duration of occupation created a confusing mass of post holes and other cut features, and it was difficult to determine in homogeneous soils the level from which features were cut. Often features observed in one trench could not be traced in the next. The fills of features were very similar to the surrounding occupation deposits, especially when dry, resulting in a lack of complete building plans.

However, a long structural sequence emerged, beginning with circular buildings of the Bronze Age, followed by a less well-defined Iron Age phase. In the Roman period, a pair of adjacent ditched enclosures was created, one of which contained several roundhouses of at least two phases. They were followed by rectilinear buildings into the 4th century and beyond, and the enclosure ditches were probably filled in during or after the 4th century. The latest occupation phases contained a large timber post-built structure, associated with a small amount of medieval 13th and 14th century pottery.

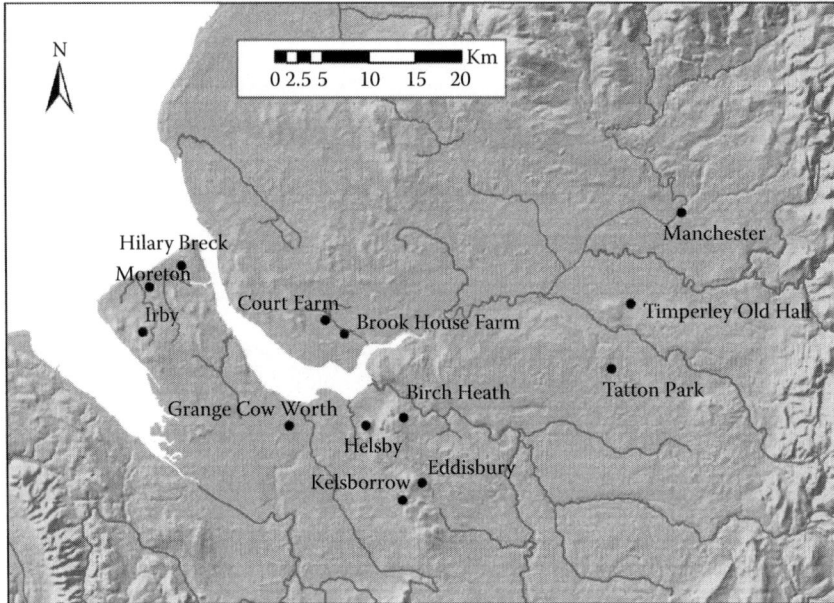

FIGURE 7.1   Locations of excavated early medieval rural sites in lowland north-west England (see Table 7.1).

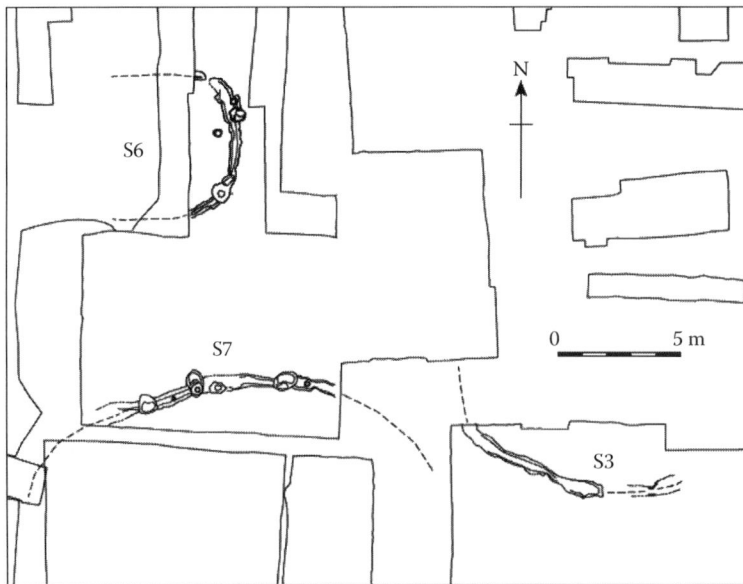

FIGURE 7.2   Irby, Wirral. Excavation plan showing three adjacent early medieval buildings, S3, S6, and S7.

An unexpected discovery in this context turned out to be rare evidence of early medieval settlement. Careful analysis of the complex stratification by Dr. Mark Adams led to the recognition of a phase of distinctive buildings. A clearly defined series of curvilinear structural gullies had cut through occupation deposits into the bedrock (Philpott and Adams 2010, pp. 52–64, Figure 2). Crucially, the stratigraphy showed that one example cut through an in-filled Romano-British ditch while another lay too close to another Romano-British ditch for the ditch and building to have been in contemporaneous use.

In a separate phase of similar date, another building, S16, was constructed not with continuous gullies but with stone rubble foundation walling, though slightly bow-sided in outline and with evidence of an internal partition. It was sealed by the clay floor of a later medieval building. The forms of these buildings with their bow-sided or outwardly curving walls are characteristic of Anglo-Scandinavian houses (Richards 2000). Evidence for such buildings is known elsewhere in the Irish Sea region (Griffiths 2010, pp. 66–71), but rarely in England.

Dating evidence was scant. Unsurprisingly, during cut through occupation deposits of a Romano-British rural settlement, all the foundation gullies were found to contain small quantities of residual Roman pottery. An important breakthrough was the identification by Julie Edwards of a spike lamp of 10th to 12th century date (Figure 4, 4). The lamp securely stratified in a building foundation gully provided the first unequivocal evidence of a structural phase of activity as early as 900 to 1200, and potentially from the pre-Conquest period. The absence of other datable artefacts along with environmental samples that lacked sufficient securely associated material to produce a reliable date meant that other direct dating evidence was lacking. The presence of a single amber bead, although not diagnostic, is consistent with the Viking period.

The tantalising structural and artefactual evidence needs to be considered with evidence of the place-name for the township. The name combines Norse elements *Íri* (Irishmen) and the suffix *-býr*, meaning farm of the Irish (Dodgson 1972, p. 264) and provides an explicit reference to Hiberno-Norse settlement within the township.

The identification of the Irby site as a Viking age settlement is thus based on several converging strands of evidence: the distinctive building form, the presence of a Saxo-Norman lamp, and the analysis of the stratigraphic sequence, combined with circumstantial place-name evidence of Hiberno-Norse settlement. If the interpretation is correct, this is the first time that archaeological evidence of a Viking rural settlement has been recognised in lowland north-west England.

There is a further implication for the township. As Old Norse *Íra-býr* (farm of the Irishmen) (Dodgson 1972, p. 264), Irby is a strong candidate for development during the primary phase of settlement of northern Wirral by a mixed group of Norse and Irish who were expelled from Dublin in AD 902 under Ingimund (see Griffiths, Chapter 2). The Old Norse settlement name that distinguished the occupants by ethnic origin rather than by personal name amongst the dominant Norse-speaking stratum of society in the Viking enclave of north Wirral clearly replaced any earlier place-name.

The fact that the farm or settlement took its name from the incomers indicates they had some pre-eminence as local landholders. In addition, the township was granted as a whole to the abbey of St. Werburgh in Chester in 1093. By that time, Irby township had already been divided between two neighbouring parishes in a complex interlocking landholding pattern that developed in the preceding century or more before being fossilised by the parish boundaries.

This division of the township may have reflected a partition between two different landowners, conceivably between the Irish incomers and an existing Anglo-Saxon owner. If so, the unnamed Irishmen who arrived in 902 and were granted land by Æthelflæd 'near Chester' (Griffiths, Chapter 2) may have been accommodated alongside the estate of an existing Saxon lord, perhaps selecting an earlier settlement site in a deliberate act of appropriation. The Irish place-names of nearby Arrowe and Noctorum townships suggest these were the location of the summer pastures attached to the 'farmstead of the Irishmen.' Both stand on a dry sandstone ridge, as reflected in the Noctorum place-name from OIr *cnocc tírim* (dry hill) (Dodgson 1972, p. 268). The Arrowe name is derived from MIr *airge* (at the sheiling) (ibid., p. 262).

Moreton

The second site at Moreton (Figure 7.3) lies 5 km north-west of Irby. Excavations were undertaken in piecemeal fashion in 1987 and 1988 after the demolition of houses at 240 to 246 Hoylake Road. Only brief references have been published (Newman 2006, p. 99; Philpott and Adams 2010, p. 212; Griffiths 2010, p. 71). A lack of funding for this rescue project from the pre-developer-funded era has meant the site report has taken many years to complete (R. Philpott in preparation). However, the importance of

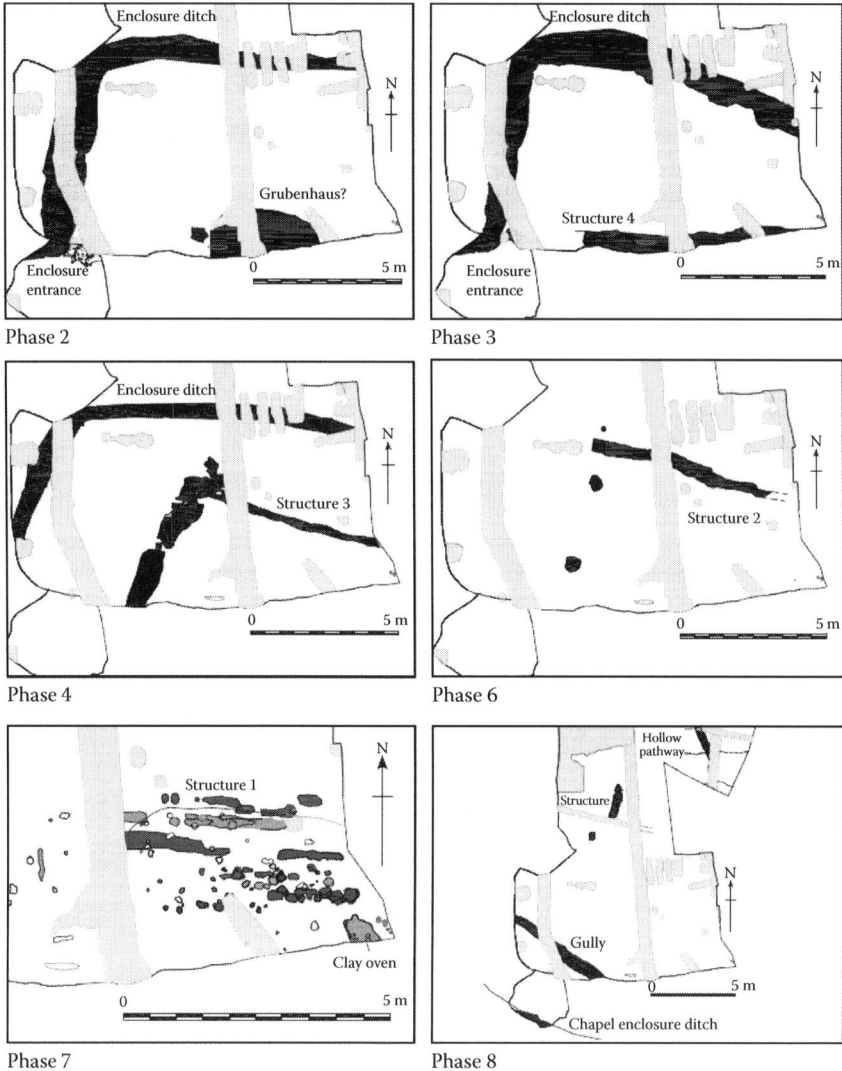

FIGURE 7.3 Moreton, Wirral. Development of Phases 2 (pre-10th century) through to 8 (14th century and later). Modern features are shown in grey.

the site was reaffirmed in 2011 when two radiocarbon dates confirmed the suspected early medieval date for much of the long stratigraphic sequence.

The excavations targeted the oval enclosure around a medieval chapel of Bidston parish. Moreton chapel was first recorded in 1347 and demolished about 1680 to 1690. No clear physical remains of the chapel were found (Figure 7.3, Phase 8) but an unrelated sequence of at least five superimposed structures was discovered within the enclosure. The earliest took the form of a large pit (S5), subrectangular in plan, measuring at least 3.5 m east–west and at least 1.5 m wide and up to 0.50 m deep although the full extent could not be determined because the southern part lay outside the excavated area (Figure 7.3, Phase 2).

The structure appears to be a *Grubenhaus*, a sunken-featured building of a type developed in the early Anglo-Saxon period (Tipper 2004). Such structures are most common in areas of Germanic settlement in the south and east of England, and are rarely found in the north-west. This appears to be the first example of the type in the historic county of Cheshire (cf. Tipper 2004, Figures 4 and 5). This type should be distinguished from the sunken-floored buildings of 10th century date found in Chester, with their perpendicular walls and absence of gable posts, and also those at Manchester, where a group of four pits was found in a line parallel to the stone wall of the later Roman fort at Northgate (Bryant et al. 1986, pp. 54–56, Figure 4.27) and at Rhuddlan, Flintshire, which David Griffiths (2010, p. 131) convincingly interpreted as associated with the early 10th century *burhs*.

Overlying the large pit at Moreton were several other structures. All the features had been heavily truncated by the plough. No complete building plans were recovered but the sequence was clear. A series of three structures was constructed, each with shallow foundation gullies. The earliest, Structure 4, survived only as a straight gully, probably the northern wall of a building, but it lacked associated structural and dating evidence (Figure 7.3, Phase 3). Overlying S4 but on a different alignment in relation to the nearby enclosure ditch, was a structure of rectangular plan (S3). Parts of two sides with dimensions of at least 4.7 m north-east by south-west by 6.5 m were recovered (Figure 7.3, Phase 4).

The northern wall was formed by a continuous foundation gully that had been deliberately packed with clean yellowish clay. The western wall was constructed with two short stretches of a shallow foundation gully separated by a post hole. A semi-circular probable post impression was observed in the base of the foundation gully. The discontinuous foundation gully and the post impression indicate the use of an interrupted post-in-trench construction—a building form characteristic of the Anglo-Saxon period.

The earliest three phases of structures were enclosed by a ditch that was recut at least twice. The north-western angle of the enclosure was identified, as was a probable entrance on the western side. Soon after the first recut of the ditch began to fill in, burnt grain was deposited near the base. The cereal grains produced a radiocarbon date of *ca.* AD 880 to 990,[*] which was consistent with the only closely datable find from the top fill of the same ditch. This was a unique silver penny of King Eadwig (955–959), minted at a south-western mint, probably Exeter, identified by Marion Archibald of the British Museum (Cook and Besly 1990, p. 229, Plate 22, Figures 1 and 4). Thus, this phase of the ditch was filled in by the mid 10th century, suggesting that the first phase of building dated to the 9th century or earlier.

After the disuse of the ditches, the next significant phase of activity saw the construction of another rectilinear building (S2). The principal element was a long shallow foundation trench, with a clearly defined squarish terminal to the west. It too was heavily truncated, surviving to a maximum depth of only 0.12 m. Two large post holes set 2.8 m apart centre to centre and at right angles to the trench terminal were most likely associated with the gully and marked the end gable wall. Together these features indicate a building with dimensions of at least 7.2 m east–west by about 5.5 m north–south. The building plan appears to be a long slightly curving rectilinear structure with individual posts supporting the end gable wall (Figure 7.3, Phase 6).

Although there are very few comparative examples in lowland north-west England, structures, S2, S3, and S4 all have parallels in Anglo-Saxon contexts elsewhere in England. Earth-fast individual post-built structures are characteristic of Anglo-Saxon timber building from the 5th century, and continued into the late Saxon period. However, plank-in-trench or post-in-trench foundations that were introduced in the late 6th century were increasingly used and represent half of the buildings in the 7th century and three-quarters by the 8th and 9th centuries (Addyman et al. 1972; Hamerow 2012, pp. 22–24).

The use of interrupted post-trench construction in which a discontinuous gully was separated by post holes is found at a number of Anglo-Saxon sites in England and southern Scotland such as Portchester Castle (S6: Cunliffe 1976) and Cheddar (Murray and Ewart 1981). At the Anglo-Saxon monastery in Hartlepool, small rectangular buildings of mid 7th to mid 8th century date included one (Building VIII) of interrupted post-trench construction in which timber posts were set vertically in a discontinuous foundation trench (Daniels 1988, pp. 172–173, Figure 22).

The final structure (Structure 1) marked a significant shift in building form (Figure 7.3, Phase 7). Structure 1 had a hollow floor, a clay-lined oven, and walls lightly constructed in wattle. The walls appeared to have been rebuilt twice on a slightly different line. The three phases matched a sequence of three slightly offset lines of stakes that presumably marked the edge of an internal low platform or bench. This suggested a division into a platform or bench along the northern wall and a central area containing the base of a single-chambered clay oven.

---

[*] Cal BP 1060 to 960; 2 sigma, Beta-309094.

The hollow floor accumulated a deposit of burnt organic material including weed seeds and cereal grains from oats, barley, and wheat, possibly free-threshing bread wheat (*Triticum aestivum*)—an identification consistent with the absence of wheat glumes or spikelets (P. Tomlinson personal communication). A new radiocarbon date from this deposit assigned the infilling to the mid 11th to early 13th century.[*]

Structure 1 has a number of features paralleled in 10th to 12th century buildings elsewhere. The plan of the building resembles those of 11th century dwellings in Dublin (Murray 1981) with their central passages, central hearths, and benches on either side.

Apart from the silver penny mentioned above, artefacts were very sparse. The total early medieval assemblage consisted of a mudstone pendant hone (Figure 4, 3) of a type found at pre-Conquest sites, a lead spindle whorl dated broadly to the 6th to 11th century (P. Walton Rogers, Figure 4, 2), and an unstratified pottery sherd that lacked close parallels and could be early medieval in date (Ailsa Mainman and Ewan Campbell, personal communications). A small quantity of poorly preserved animal bone was subject to a preliminary scan by Dr. Sue Stallibrass, who reported horse, cattle, sheep, and pig.

Hints of earlier occupation were found in the immediate vicinity. A Byzantine coin of Justinian I minted in Carthage in AD 540 was found in a garden less than 300 m away (Philpott 1999). A suggestion of Roman activity nearby comes in the shape of a probable residual sherd of Roman pottery and a 4th century coin on the site, indicating that the Anglo-Saxon settlement probably shifted to a new site near the earlier occupation by the 9th century.

The chief significance of Moreton is that the building forms appear to remain Anglo-Saxon in character, as far as we can tell from a region where such structural evidence is very sparse, from its beginning at least by the 9th century until the mid 11th century. One interpretation is that the occupants continued to build in an Anglo-Saxon tradition, only to be replaced in the 11th century, perhaps by incomers who constructed in a style then current in Dublin and also present elsewhere, in Whithorn and York, for example (Murray 1981 and 1983; Hill 1997, pp. 209–217).

If, as may be hypothesised, this indicates a takeover of an Anglo-Saxon estate by a Norse lord, as it appears at face value, then it occurred fairly late and several generations after the primary Viking settlement in the area. The construction may alternatively be a reflection of changing building fashions across ethnic communities, but it suggests the adoption of building forms familiar from a Scandinavian milieu, whether from the Irish Sea region, Ireland, or the mainland of England or Scotland. The proximity of the port of Meols and its finds assemblage containing Hiberno-Norse style objects into the 11th century may demonstrate the direction from which ideas were transmitted by an individual of Norse ancestry or affiliation (Griffiths, Chapter 2).

Locating Viking Age Settlements

Material Culture

The two Wirral sites discussed above highlight some of the key difficulties in identifying early medieval sites in the region (Figure 7.4). To assist, we may enlist the help of studies from other regions to provide a methodology.

Recent artefact analyses have begun to unlock the vast potential of the finds recorded by the Portable Antiquities Scheme (PAS) in northern and eastern England. Studies of ornamental metalwork such as brooches and strap ends have recognised distinctions between Scandinavian pieces that are likely to have been brought from the Scandinavian homelands by settlers or traders and Anglo-Scandinavian items that represent interactions between Anglo-Saxons and Scandinavians (Thomas 2000; Kershaw 2009).

The research allowed the identification of those within the Danelaw who used material culture to express Scandinavian cultural affiliations (Kershaw 2009). The presence of distinctive Hiberno-Norse or Scandinavian objects such as ringed pins, weaponry, and other ornamental metalwork offers the potential to identify Scandinavian settlements or contacts around the Irish Sea, as demonstrated by the items

---

[*] Two sigma calibration results of Cal AD 1040 to 1110 (Cal BP 910 to 840) and Cal AD 1120 to 1220 (Cal BP 840 to 730). Beta 307737: 880 ± 30 BP to 24.3‰, 890 ± 30 BP.

FIGURE 7.4 Early medieval finds assemblages in Wirral. Moreton: 1. Coin of Eadwig. 2. Spindle whorl. 3. Hone. Irby: 4. Saxo-Norman spike lamp.

buried in the small family cemetery for first or second generation settlers at Cumwhitton, Cumbria (Newman 2009).

In a valuable series of articles, Julian D. Richards (1999, 2000, 2001) considers how to use archaeological evidence to identify Anglo-Scandinavian rural settlements. Using the example of Cottam in the East Yorkshire Wolds, he sets out three main criteria for the identification of an Anglo-Scandinavian rural settlement in the northern Danelaw: (1) discontinuity of settlement, (2) culturally distinct artefact types, and (3) distinctive building forms. Slight evidence for distinctive building forms has been considered above. To what extent can we apply the other criteria to the sites in north-west England?

Richards argues that one of the key elements in distinguishing an Anglo-Scandinavian settlement is the identification of finds. He warns that we cannot claim that 'the inhabitant of a settlement was Scandinavian' but suggests that we can 'identify the use of material culture to proclaim a particular identity, and to invent new ones' (2000, pp. 302–303). The 'Anglo-Scandinavian' term denotes the creation of a new identity 'arising from cultures in contact.'

Field walking and metal detecting at Cottam produced two distinct concentrations of finds. Objects of 8th to 9th century date (no fewer than 68 dress pins, 34 strap ends, 35 knife blades, 8 lead weights, other objects, 3 *sceattas* from the 8th century, 22 *stycas* from the 9th century) occurred largely in one area. A concentration of Anglo-Scandinavian metal finds, including weights, two 'Norse' bells, objects decorated in Scandinavian styles, and 10th century Torksey ware was found in another area. After plotting the features through geophysical survey and excavation, the finds could be seen to be associated with two discrete crop mark enclosures. An Anglian settlement associated with the earlier finds was abandoned and replaced by an early 10th century Anglo-Scandinavian farmstead on a separate site to the north-east.

In theory, the application of Richards's suite of techniques may appear to offer an approach to the problem of site location. However, marked regional differences mark the north-east and north-west of England. The type of rich material assemblage found at Cottam is not present at rural sites in north-west England. The region appears virtually aceramic from the 5th to 12th centuries except for towns such as Chester where late Saxon Chester ware appears from the early 10th to mid 11th centuries (Carrington 1977; Axworthy-Rutter 1985 and 1994).

There is of course a danger of engaging in a circular argument that Chester Ware is mostly found in urban contexts and rarely at rural sites and is largely an urban phenomenon. Few rural sites have been located and even fewer extensively excavated so that the absence of such items at rural sites has

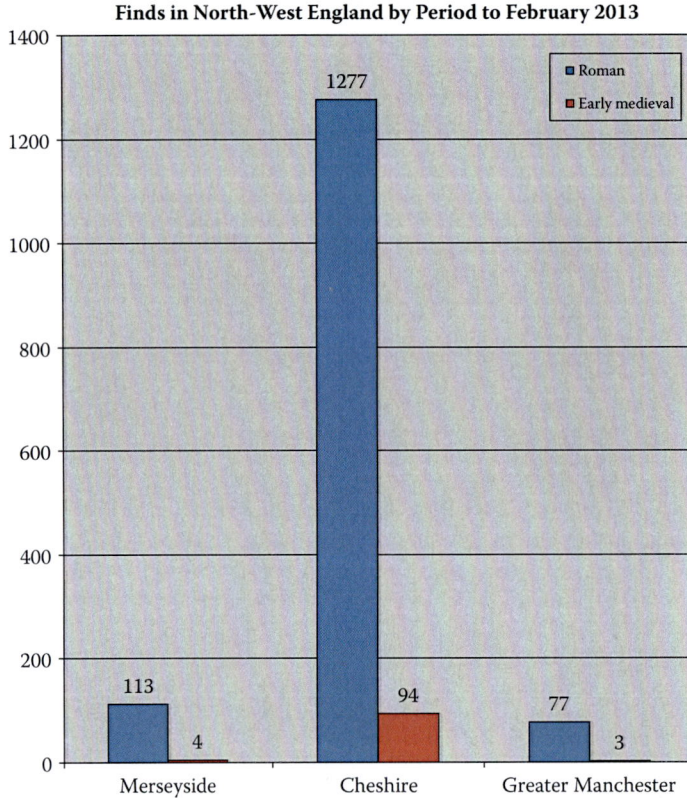

Finds in North-West England by Period to February 2013

FIGURE 7.5 Post-1974 finds of Roman and early medieval date in lowland north-west England reported to Portable Antiquities Scheme.

not been proven. One consequence of the rarity of ceramics is that it is impossible in the north-western counties of England to undertake a study such as the Currently Occupied Rural Settlements (CORS) survey in Cambridgeshire and Norfolk (Thomas 2012, p. 33) that uses distributions of datable pottery to identify shifts in settlement nuclei between the mid and late Saxon periods (Lewis 2007). In the virtual absence of early medieval pottery, the effectiveness of field walking is restricted for locating sites since it relies on locating diagnostic artefacts with sufficient frequency to identify spatial distributions and concentrations.

Metal finds are a little more forthcoming but the only site in the region to produce a relatively large quantity of early medieval finds is the beach market and trading settlement at Meols, which is unusual in its circumstances of loss, burial, and retrieval (Griffiths, Philpott and Egan 2007; Griffiths, Chapter 2). However, significant finds assemblages from rural settlement sites have not been forthcoming in north-west England. More typical is the pattern represented by the Portable Antiquities Scheme (PAS) showing that reported early medieval finds consisting mostly of metalwork are not common.

There are only 101 early medieval finds for Greater Manchester, Cheshire and Merseyside, compared with 1467 Roman finds recorded to February 2013 (Figure 7.5), most of which come from Cheshire. Another measure is the national picture of early medieval finds in the north-west compared with other regions. Rural settlements in northern and western England and Wales present a stark contrast with those in the south and east.

The Viking and Anglo-Saxon Landscape and Economy (VASLE) project[*] that plotted all PAS finds known to 2006 to create a map of find spot densities across England and Wales confirmed what had

---

[*] VASLE was a 3 year project to investigate coin and artefact data for the period 700–1050 for spatial distributions and study settlement hierarchies and the development and settlement morphologies of specific sites (Richards and Naylor 2008).

previously been a consistent yet unquantified impression. Marked regional disparaties are evident with a strong emphasis on early medieval finds in the south and east of England, and particular concentrations in East Anglia, Kent, North Lincolnshire, parts of Yorkshire, and Northamptonshire (Richards and Naylor 2008, pp. 173, Figure 15.3). Elsewhere significant gaps exist, notably in north-west England, South Wales, the Midlands, and the south-west of England, all with very low find densities (Richards and Naylor 2008, pp. 173–175, Figure 15.4).

The extent of regional disparity is illustrated by the Yorkshire Cottam site (Richards 2001). Richards (1999, p. 1) observed 'the results suggest that it was a prosperous but not exceptional site, and that the primary activity was farming, with limited evidence for trade or manufacture,' yet the total for this site alone exceeds the number of early medieval finds recorded by the PAS through early 2013 for the north-western counties of Cheshire, Merseyside, and Greater Manchester, with a combined area of 4264 km².

We are confronted therefore by a significant regional difference between material culture uses in the west and east that impact our ability to locate settlements. Thus, while the use of diagnostic metal methods is an important tool in site location, the limited use recorded to date suggests that metal detecting may locate few settlement sites of the period.

Even when characteristic finds are made we know their precise find spots; finds alone are not necessarily good indicators of settlement locations. Items such as the Viking silver ingot from South Wirral, a lead weight from Saighton near Chester, and the few recorded coins are all highly portable and easily lost. As for hoards, although four Viking-age hoards have been recovered from Chester (Edwards 1998, pp. 51–52), the place of concealment may have been located deliberately away from settlements, as may well be the case for Cuerdale on the bank of the Ribble (Graham-Campbell 2011) and Eccleston and Huxley (Graham-Campbell and Philpott 2009).

Durable material culture of early medieval date at the few excavated sites has also proven very scarce. Early medieval sites are extremely difficult to recognise in the first place and after they are identified, it is very difficult to establish a dated sequence of activities. That this is not a problem unique to the northwest of England is illustrated by Hamerow's comment from a national perspective that 'most Anglo-Saxon rural settlements are … disappointingly "clean" in archaeological terms, yielding few finds other than pottery and bone, and many producing precious little even of these' (2012, p. 2).

Compounding the difficulty of identifying sites is recognising early medieval phases when they *are* present at excavated sites. A number of excavations on rural settlements revealed unexpected evidence of early medieval occupation, usually relying on radiocarbon dating to distinct early medieval from the dominant visible Romano-British or later medieval occupation. At other sites, the complete absence of datable finds, particularly pottery, allied with particular structural forms leads only to a suspicion that an early medieval phase is present.

Such is the case with a heavily truncated structure partially excavated on Hilbre, a small tidal island lying off the Wirral coast at the mouth of the Dee Estuary by the writer in 2006 and 2007. A row of rock-cut post holes, probably forming part of one wall of a rectilinear building of earth-fast posts and associated with a hearth lacked any datable finds. There was a historically-recorded early medieval chapel or monastic cell on Hilbre (Griffiths, Philpott and Egan 2007, pp. 369–371).

During the mid 19th century, a small burial ground, apparently dating to the Early Christian period that produced an Anglo-Saxon glass bead was noted by local antiquarian Henry Ecroyd Smith. The site was located near the centre of the small island, in an area that also produced a sandstone cross of Viking type and a grave slab, both dated to the 10th or 11th centuries and probably associated with the chapel (Bailey 2010, pp. 81–82). The 'island of Hildeburg' place-name may refer to an otherwise unrecorded Anglo-Saxon saint (Dodgson 1972, pp. 303–304; Thacker 1987, p. 269). Potential exists for early medieval and Viking Age activities on Hilbre but without radiocarbon dates we have no definitive evidence.

Furthermore, in the absence of diagnostic objects, we cannot use the material culture to attempt to make statements, however imperfect or ambiguous, about the ethnicity, status, or identity of the occupants. Nor can we identify the kinds of changes that may be observed in areas of richer material culture such as western Scotland where materials of native origin were supplemented or replaced by objects with Scandinavian associations or the Danelaw where distinctive metalwork was used to forge and express Scandinavian or Anglo-Scandinavian identities.

Re-Occupation of Settlement Sites

The final criterion set out by Richards for recognition of Viking rural settlements is discontinuity of settlement based on a shift in location between Anglian and Anglo-Scandinavian phases at Cottam. Here we may also find that the distinctive characteristics of settlement patterns in north-west England do not make it easy to recognise the phenomenon of settlement shift.

Higham (2004a, p. 23) notes the accidental nature of the identification of early medieval sites at Tatton and Manchester, both of which were investigated for sites of other periods. It is striking how dependent we appear to be on chance for location of sites. Sites are not located by mere chance, illustrating a strong degree of conservatism in the location of settlements in lowland north-west England. Intermittent settlement, repeatedly re-occupying the same site, sometimes after centuries of abandonment, is a consistent phenomenon at rural sites. Often these re-occupations involve the re-use of earlier enclosures that survived as earthworks into the early medieval period.

Reuse of a substantial Iron Age double-ditched enclosure at Brook House Farm, Halewood was demonstrated by radiocarbon dates from the infilling of the massive enclosure ditch that indicated activities in the 11th to 13th centuries (*ca.* 1000–1230: Cowell and Philpott 2000, p. 65). Although it was considered most likely associated more with clearance for agriculture than with settlement, a number of unphased and undated buildings lacking diagnostic finds in the enclosure interior may well belong to the same phase of activity. It remains a distinct possibility that re-occupation of the enclosure occurred during the late Anglo-Saxon period and the settlers utilized the enclosure ditches that survived as substantial hollows.

The reuse of Iron Age hillforts or hilltop enclosures such as at Kelsborrow Castle (Garner 2012, p. 67) suggests that a characteristic of the late Anglo-Saxon period was re-occupation of prominent sites, firmly embedded in the landscape and defined or protected by substantial earthworks. A defensive role may have been part of the rationale in selecting these sites, as in the foundation of the Æthelflædan *burh* of AD 914 on the Iron Age hillfort at Eddisbury (Garner 2012, p. 66).

Underlying this may be a growing concern to demonstrate rights of landownership at a time of expanding population, increasing the competition for land and needs to differentiate rank and status. Because not all of the re-occupied sites have substantial earthworks, there may be an emphasis as much on the consolidation of rights over land, perhaps through appropriation of former ancestral settlements. Sites of the latter type include the Romano-British enclosure at Irby, where the ditches were infilled by the late Roman period, although an undated phase of ditches suggests they were redefined in the early medieval period.

Moreton appears to be slightly different, with the creation of fairly insubstantial enclosure ditches during the early medieval period, although they had gone out of use by Phase 5, dating probably to the late 10th or 11th century. This raises a question about the dating of some of the dozens of discrete enclosures noted as cropmarks through aerial reconnaissance in Merseyside, Cheshire, and West Lancashire since the 1970s. Despite the difficulties of examining expanses of glacial till, clay soils, and extensive pastures that are rarely conducive to the formation of cropmarks, aerial reconnaissance at best is most effective in this region for identifying the distinctive signatures of sites enclosed by ditches (Collens 1994).

The enclosures usually produce late prehistoric or Romano-British finds or both, but we can rarely be confident that they are *exclusively* indicative of late prehistoric or Romano-British occupation. The cores of some original Anglo-Saxon or Anglo-Scandinavian settlements may have been defined by ditched enclosures. We know so little about early medieval settlements that it is not yet possible to recognise their diagnostic cropmark signatures on morphological grounds. In examining aerial photographic evidence therefore, we must be open to the possibility that some enclosures may be early medieval or Viking Age settlements.

The soils and landscape of north-west England undoubtedly present methodological challenges to aerial reconnaissance. However, one consistent landscape characteristic revealed by aerial photography is the rarity of field systems that can be shown to pre-date the earliest historic maps. Even in the most productive years, aerial reconnaissance failed to identify extensive field and enclosure systems characteristic of large areas of southern and eastern England. The creation in the middle Saxon period of more bounded landscapes, characterised by systems of ditched enclosures to demarcate and regulate the landscape has been seen as evidence of social transformation, reinforcing growing differences in rank in society (Hamerow 2012, pp. 98–99).

The creation of extensive bounded landscapes appears to form a marked regional contrast to the settlement pattern of dispersed farms and hamlets in the less densely populated north-west of England where long-term stability of settlement and landscape elements is argued by Atkin (1985) and others. A different picture of settlement location may be tentatively emerging. Historic maps record a number of instances in Lancashire and Cheshire where field boundaries appear to have fossilised large curvilinear enclosures, often encircled by trackways and abutted by other boundaries that appear to represent primary agricultural enclosures.

The existence of early defined blocks of fields preserved as curvilinear, often oval, patterns was noted at Tarbock and Lathom (Cowell and Philpott 2000, pp. 208–210). Some were certainly in use in the medieval period (Hunsterson and Cheshire: Roberts and Wrathmell 2002, pp. 99–100, Figure 4.12; Atkin 1985; Annakin-Smith 2012). However, earlier origins are possible in the light of the existence indicated for a morphologically very similar field pattern at the Romano-British settlement of Roystone Grange, Derbyshire (Hodges 1991) where one of the paired enclosures was used for arable purposes, the other for pasture.

Excavation shows some enclosures were associated with long-term settlement. At Brunt Boggart in Tarbock, the occupation was intermittent rather than continuous, first in the Bronze Age, followed by Roman, and after a long interval by later medieval phases. The same location adjacent to the oval enclosure was reused for settlement. In other cases, the enclosures may have been long-lived features of the agrarian landscape, but not all are necessarily pre-medieval in origin. Independent or irrespective of date, the oval form has been seen as the 'primary agrarian structure, unconstrained by any pre-existing boundaries in a particular locality' (Roberts and Wrathmell 2002, p. 114).

The shape represents the most economical form in terms of effort of setting out enclosures in a newly laid out landscape. The process of developing 'loop forms' of enclosure was illustrated by Roberts and Wrathmell (2002, Figures 6.3 and 6.4). 'Ring-fenced enclosures' normally formed the foci of important farmsteads, halls, and even townships and parishes and may be of considerable antiquity, pre-dating the medieval period (Roberts and Wrathmell 2002, pp. 152, 163–164).

Cultivated land was cleared at considerable expense. Extensive long-term effort was needed to cut woodland and remove stumps, clear surface stones, dig ditches for drainage, and maintain soil fertility through manuring. Further effort was required to construct and maintain boundaries, whether banks, ditches, or hedges. Roberts and Wrathmell determined that breaking and ploughing an acre of land required a ploughman to walk 11 miles (2002, p. 40). There was no sweeping away of the previous landscape organisation with the arrival of the Anglo-Saxons or Vikings (Higham 2004). In a landscape that was sparsely populated, the core areas of arable land are likely to have remained stable, embedded within the landscape where they often occupied favourable niches.

The effort of defining and maintaining the investment of an early core of arable land meant that settlement periodically moved around it, exploiting the same core land but occupying and re-occupying the same settlement sites within estate boundaries, sometimes after several centuries or longer. Repeated occupation of the same site sometimes after long intervals is a potentially useful phenomenon that enables us to identify sites. However, it will only work for sites that remained in use or were re-occupied; it will not help locate sites where occupation was curtailed or short-lived.

Place-Names

Place-names are not usually attached to points. They are more likely given to areas of land, agricultural units, or estates and named settlements may shift within a given area. In Wirral, if correctly identified, the Irby place-name refers to an excavated site over 800 m from the later medieval village nucleus.

The mobility of settlements in the early medieval period is well established (cf. Taylor 1983; Hamerow 2012, pp. 67–70; Thomas 2012, pp. 44–45). We cannot be sure that a named place is in the same location as a later village or farm since settlements may have shifted locations. The 'drift' concept identified by Chris Taylor (1983, pp. 120–123; Roberts and Wrathmell 2002) is seen widely across southern and eastern England in the early to mid Anglo-Saxon period.

Place-names provide some clues to the locations of sites, especially where they are preserved in field names and localised to a relatively small area. Dodgson identified a series of habitative place-names that may represent 'lost' settlements of possible Scandinavian origin in Wirral and nearby. They include Haby in Barnston, Hesby in Bidston, Warmby in Heswall, Kiln Walby in Overchurch (Upton), Stromby in Thurstaston, and Syllaby in Great Saughall (Dodgson 1972, *passim*).

There is growing agreement amongst place-name scholars that -*by* names represent low-status places, less substantial than names ending in the Old English –*tūn*, and are located in relatively marginal places away from major centres of power (Abrams and Parsons 2004, p. 401). The Wirral sites bear out this interpretation as secondary, minor places within the framework of English parish and township names, as several lie in marginal locations near township boundaries.

The value of place-names as indicators of early medieval settlements can be seen in recent work at Wallasey. The site at Hilary Breck is the only early medieval site discovered so far as a result of fieldwork specifically targeted for that period. The site lay on a slope just below the hilltop parish church, with its early dedication to St. Hilary in Wallasey at a place formerly known by the Norse place name Kirkby in Waley (ON 'church-village,' with parish name Waley, 'the Britons' island': Dodgson 1972, pp. 324, 332). National Museums Liverpool's Archaeology Department found a corn-drying oven associated with a deposit rich in cereals and other carbonised plant material (Adams 2012). Radiocarbon confirmed a very late 4th to 6th century AD date, ca. 390 to 580. AD indicating activity from the highly elusive early post-Roman period. Although not Viking Age, the place-name provided crucial evidence of the potential for early medieval activity.

Gill Chitty suggested that the influence of the Scandinavian settlers on the landscape can be seen in the small townships of Arrowe, Pensby, and Noctorum, all occupying spurs of elevated land with single farm holdings (1978, p. 10) that may indicate the colonisation of previously unoccupied land for pastoral farming by Norse settlers. She further suggested that air and field surveys might produce evidence at Hargrave, a small Domesday manor with a Norse-named Lord Osgot at Domesday. The manor later was incorporated into Raby. Was there ever a nucleated settlement there or was it always occupied by a single farmstead (Chitty 1978, p. 10)? Such a site, largely unencumbered by destructive later development, may offer good opportunities to identify early medieval occupations.

Conclusions

The identification of early medieval settlements in lowland north-west England remains a formidable challenge. To make progress in site location we require an increasingly sophisticated and sensitive approach to fieldwork, employing a combination of several techniques. Against the somewhat pessimistic assessment of the potential for material culture to assist with site location, some indicators can help narrow down the locations of early medieval sites. Detailed studies of the place-names allied to the examination of field systems and settlements may help examine the options.

More sophisticated use of geophysics aided by technical advances to target archaeologically ephemeral sites needs to be allied to other field techniques. Field-walking strategies need to pay close attention to undiagnostic cultural material such as concentrations of fired clay that, in the absence of datable artefacts such as Roman pottery, may be early medieval in date and areas of former occupation where the dark soil is enriched by organic material or charcoal.

Similar indicators should be observed in topsoil stripping in advance of development where the ground has not been ploughed. Indicators such as find spots, particularly where more than one artefact is recorded, deserve close scrutiny. The vicinity of the find spots of early medieval metalwork or coins may benefit from field walking, in case they turn out to be indicators of settlement rather than isolated chance losses. The use of hand-held GPS technology in field walking makes it relatively quick and easy to plot accurately changes in soil types and find spots.

Metal detector users are encouraged to report accurate find spots (ideally 12-figure grid references, accurate to the metre), again supported by GPS technology that is increasingly available on mobile phones and on detectors. Precise locational data will allow the identification of concentrations of finds that over time may allow recognition of settlements as opposed to single stray finds, and even enable distinct phases of settlement shift to be identified. Another technique is combining different data sets through GIS to correlate find spots with sites identified through place-names or field-names and crop marks to reveal patterns or concentrations of potential significance.

An important methodological issue has implications for the way large area developments are treated in the planning system. As Gabor Thomas observes, commercial developments have begun to populate the landscape but only where extensive areas have been examined for roads, housing, and other schemes (2012, p. 44). Without extensive stripping for archaeological purposes of large areas of land, we may not be sure we have located all the settlement evidence.

An excellent example is Court Farm, Halewood where an extensive Romano-British rural settlement, a hamlet, was abandoned, only to be reoccupied in the 8th and 9th centuries (Adams and Philpott forthcoming). The Romano-British occupation was identified only through field walking because a large housing development was thought (incorrectly, as it transpired) to have a moated site.

A further important consideration is that excavators, and especially contracting archaeologists from other regions who may not be familiar with the ephemeral remains and methodological issues, need to be aware that sites dominated by relatively material-rich periods such as the Romano-British or late medieval may have early medieval phases lurking amongst undated features, so they may actively seek funding for radiocarbon dates for apparently aceramic features or phases. In the end, extensive topsoil stripping, as at Hilary Breck in Wallasey, may be the only secure way to ensure that ephemeral sites poor in material cultural remains will be identified.

In the early 1980s, very few Romano-British rural sites were known in lowland north-west England. Concerted research programmes and the application of improved field techniques have considerably advanced our understanding for that period. Whilst acknowledging the severe limitations of the evidence, the application of improved methodologies may enable us to narrow down the search for early medieval settlements. This chapter has ranged well beyond the Viking Age but the problems of the settlement location are common to the wider early medieval period. It is time to apply new approaches to the early medieval settlement of north-west England to transform our understanding of settlement in that period.

REFERENCES

Abrams L. and Parsons D.N. (2004). Place-Names and the History of Scandinavian Settlement in England. In Hines J., Lane A., and Redknap M., Eds., *Land, Sea, and Home: Proceedings of Conference on Viking-Period Settlement at Cardiff, July 2001*. Leeds, Maney, pp. 379–431.

Adams M. (2012). An Archaeological Watching Brief on Land at Hilary Breck, Wallasey, Wirral, Merseyside. Unpublished report for LSP Developments, National Museums Liverpool.

Adams M.H. and Philpott R.A. *Excavations on a Romano-British and Anglo-Saxon Settlement at Court Farm, Halewood, Merseyside 1997-8,* National Museums Liverpool (forthcoming).

Addyman P.V., Leigh D., and Hughes M.J. (1972). Anglo-Saxon Houses at Chalton, Hampshire. *Medieval Archaeology* 16, 13–32.

Annakin-Smith A. (2012). Curvilinear Enclosures in the Cheshire Landscape. In Varey S.M. and White G.J., Eds., *Landscape History Discoveries in the North West*. Chester, University of Chester Press.

Atkin M.A. (1985). Some Settlement Patterns in Lancashire. In Hooke D., Ed., *Medieval Villages*. Oxford University Committee for Archaeology Monograph 5, Oxford, pp. 171–185.

Axworthy-Rutter J.A. (1985). The Pottery. In Mason D.J. P., Ed., *Excavations at Chester. 26-42 Lower Bridge Street. The Dark Age and Saxon Periods*. Chester, Chester City Council, pp. 40–60.

Axworthy-Rutter J.A. (1994). Pottery, In Ward S., Ed., *Excavations at Chester. Saxon Occupation within the Roman Fortress, Sites Excavated 1971-1981*. Chester, Chester City Council, pp. 85–91.

Bailey R.N. (2010). *Corpus of Anglo-Saxon Stone Sculpture. Vol IX: Cheshire and Lancashire*. Oxford, Oxford University Press.

Bailey R.N. and Whalley J. (2006). A Miniature Viking Age Hogback from the Wirral. *Antiquaries Journal* 86, 345–356.

Bean S.C. (2000). Silver Ingot from Ness, Wirral. In Cavill P., Harding S.E. and Jesch J., Eds., *Wirral and its Viking Heritage*. Nottingham, English Place-Name Society, pp. 17–18.

Brotherton-Radcliffe E.H. (1975). Excavations at Grange Cow Worth, Ellesmere Port, 1966 and 1967. *Journal of the Chester Archaeological Society* 58, 69–80.

Bryant S., Morris M., and Walker J.S.F. (1986). *Roman Manchester: A Frontier Settlement*. Manchester, Greater Manchester Archaeological Unit.

Carrington P. (1977). Chester: Crook Street (1973-4) Pit Group (CHE/CRS). In Davey P.J., Ed., *Medieval Pottery from Excavations in the North West*. Liverpool, Institute of Extension Studies, pp. 14–17.

Chitty G. (1978). Wirral Rural Fringes Survey. *Journal of Merseyside Archaeological Society* 2, 1–25.

Collens J. (1994). Recent Discoveries from the Air in Cheshire. In Carrington P., Ed., *From Flints to Flower Pots: Current Research in the Dee–Mersey Region* Archaeological Service Occasional Paper 2, Chester, Chester City Council, pp. 19–25.

Cook B.J. and Besly E.M. (1990). Coin Register 1989. *British Numismatic Journal* 59, 221–233.

Cowell R.W. and Philpott, R.A. (2000). *Prehistoric, Roman and Medieval Excavations in the Lowlands of North-West England: Excavations along the Line of the A5300 in Tarbock and Halewood*. Liverpool, National Museums and Galleries on Merseyside.

Cunliffe B. (1976). *Excavations at Portchester Castle. Volume II: Saxon*. London, Society of Antiquaries.

Daniels R. (1988). The Anglo-Saxon Monastery at Church Close, Hartlepool, Cleveland. *Archaeological Journal* 145, 158–210.

Dodgson J. McN. (1972). *The Place-Names of Cheshire. Part IV: The Place-Names of Broxton Hundred and Wirral Hundred*. Cambridge, English Place-Name Society.

Edwards B.J.N. (1998). *Vikings in North-West England: The Artifacts*. Lancaster, Centre for North-West Regional Studies, University of Lancaster.

Fairburn N. (2002). Birch Heath, Tarporley: Excavation of a Rural Romano-British Settlement. *Journal of the Chester Archaeological Society* 77, 58–114.

Faull M., Ed. (1984). *Studies in Late Anglo-Saxon Settlement*. Oxford, Department for External Studies, Oxford University.

Garner D. (2012). *Hillforts of the Cheshire Sandstone Ridge*. Chester, Cheshire West and Chester Council.

Gelling M. (1995). Scandinavian settlement in Cheshire: Evidence of Place-Names. In Crawford B., Ed., *Scandinavian Settlement in Northern Britain*. London, Leicester University Press, pp. 187–194.

Graham-Campbell J. (2011). *The Cuerdale Hoard and Related Viking Age Silver and Gold from Britain and Ireland in the British Museum*. London, British Museum.

Graham-Campbell J. and Philpott R., Eds. (2009). *The Huxley Viking Hoard: Scandinavian Settlement in the North West*. Liverpool, National Museums Liverpool.

Griffiths D.W., Philpott R.A., and Egan G. (2007). Meols: *The Archaeology of the North Wirral Coast*, Oxford University School of Archaeology Monograph 68, University of Oxford.

Griffiths D. (2010). *Vikings of the Irish Sea*. Stroud, History Press.

Hamerow H. (2012). *Rural Settlements and Society in Anglo-Saxon England*. Oxford, Oxford University Press.

Higham N.J. (1999). The Tatton Park Project, Part 2. *Journal of the Chester Archaeological Society* 75, 61–133.

Higham N.J. (2004a). *A Frontier Landscape: The North West in the Middle Ages*. Bollington, Windgather Press.

Higham N.J. (2004b). Viking Age Settlement in the North-Western Countryside: Lifting the Veil? In Hines J., Lane A., and Redknap M., Eds., *Land, Sea, and Home. Proceedings of Conference on Viking Period Settlement at Cardiff, July 2001*. Leeds, Maney, pp. 297–311.

Hill P. (1997). *Whithorn and St. Ninian. The Excavation of a Monastic Town, 1984–91*. Stroud: Sutton Publishing.

Jordan D. (2007). *Evaluating Aggregate in North West England: The Effectiveness of Geophysical Survey*, PDF accessed on Internet.

Kershaw J.F. (2009). Culture and Gender in the Danelaw: Scandinavian and Anglo-Scandinavian Brooches. *Viking and Medieval Scandinavia* 5, 295–325.

Lewis C. (2007). New Avenues for the Investigation of Currently Occupied Rural Settlements. *Medieval Archaeology* 51, 133–163.

Losco-Bradley S. and Wheeler H.M. (1984). Anglo-Saxon Settlement in the Trent Valley: Some Aspects. In Faull M., Ed., *Studies in Late Anglo-Saxon Settlement*. Oxford, Oxford University Department for External Studies, pp. 101–114.

Murray H. (1981). Houses and Other Structures from the Dublin Excavations, 1962–1976; A Summary. In Bekker-Nielsen H. et al., Eds., *Proceedings of Eighth Viking Congress, Århus 24–31 August 1977.* Odense, Odense University Press, pp. 43–68.

Murray H. (1983). *Viking and Early Medieval Buildings in Dublin. A Study of the Buildings Excavated under the Direction of A. B. Ó Ríordáin in High Street, Winetavern Street, and Christchurch Place, Dublin, 1962–63, 1967–76.* Oxford, BAR Brit Series 119.

Nevell M. (1997). *The Archaeology of Trafford; A Study of the Origins of Community in North-West England Before 1900.* Manchester: Greater Manchester Archaeological Unit, University of Manchester Archaeological Unit, and Trafford Metropolitan Borough Council.

Newman R. (2006). The Early Medieval Period Resource Assessment. In Brennand M., Ed., *The Archaeology of North-West England: An Archaeological Research Framework for North-West England. Vol. 1, Resource Assessment.* Manchester, ALGAO, English Heritage, and Council for British Archaeology North-West, pp. 91–114.

Newman R. (2009). The Cumwhitton Viking Cemetery. In Graham-Campbell J. and Philpott R., Eds., *The Huxley Viking Hoard: Scandinavian Settlement in the North-West.* Liverpool, National Museums Liverpool, pp. 21–23.

Parsons D. (2001). How Long Did the Scandinavian Language Survive in England? Again. In Graham-Campbell J. et al., Eds., *Vikings and the Danelaw. Select Papers from Proceedings of 13th Viking Congress, Nottingham and York, 21–30 August 1997,* Oxford, Oxbow, pp. 299–312.

Philpott R.A. (1999). Three Byzantine Coins Found near the North Wirral Coast in Merseyside. *Transactions of the Historic Society of Lancashire and Cheshire* 148, 197–202.

Philpott R.A. and Adams M.H. (2010). *Irby, Wirral: Excavations on a Late Prehistoric, Romano-British and Medieval Site, 1987–96.* Liverpool, National Museums Liverpool.

Richards J.D. (1999). Cottam: An Anglo-Scandinavian Settlement on the Yorkshire Wolds. *Archaeology Journal* 156, 1–111.

Richards J.D. (2000). Identifying Anglo-Scandinavian Settlements. In Hadley D.M. and Richards J.D., Eds., *Cultures in Contact: Scandinavian Settlement in England in the 9th and 10th Centuries.* Turnhout, Brepols, pp. 295–309.

Richards J.D. (2001). Finding the Vikings: Search for Anglo-Scandinavian Rural settlement in the Northern Danelaw. In Graham-Campbell J. et al., Eds., *Vikings and the Danelaw: Select Papers from Proceedings of 13th Viking Congress, Nottingham and York, 21–-30 August 1997.* Oxford, Oxbow, pp. 269–277.

Richards J.D. and Naylor J. (2008). The Real Value of Buried Treasure. VASLE: The Viking and Anglo-Saxon Landscape and Economy Project. In Thomas S. and Stone P.G., Eds., *Metal Detecting and Archaeology.* Woodbridge, Boydell Press, pp. 167–179.

Roberts B.K. and Wrathmell S. (2002). *Region and Place: A Study of English Rural Settlement.* London, English Heritage.

Sawyer P.H. (1976). Introduction: Early Medieval English Settlement. In Sawyer P.H., Ed., *Medieval Settlement: Continuity and Change.* London, Edward Arnold, pp. 1–7.

Taylor C. (1983). *Village and Farmstead: A History of Rural Settlement in England.* London, George Philip.

Thacker A.T. (1987). Anglo-Saxon Cheshire. In Harris B.E. and Thacker A.T., Eds., *A History of the County of Chester, 1.* Oxford, Oxford University Press, University of London, Institute of Historical Research, pp. 237–292.

Thomas G. (2000). Anglo-Scandinavian Metalwork from the Danelaw: Exploring Social and Cultural Interaction. In Hadley D.M. and Richards J.D., Eds., *Cultures in Contact: Scandinavian Settlement in England in the 9th and 10th Centuries.* Turnhout, Brepols, pp. 237–255.

Thomas G. (2012). The Prehistory of Medieval Farms and Villages: From Saxons to Scandinavians. In Christie N. and Stamper P., Eds., *Medieval Rural Settlement. Britain and Ireland AD 800–1600.* Oxford, Windgather Press, pp. 43–62.

Tipper J. (2004). *The Grubenhaus in Anglo-Saxon England.* Yedingham, Landscape Research Centre.

Wainwright F.T. (1948). Ingimund's Invasion. *English Historical Review* 63, 145–169. Reprinted in Finberg, H., Ed., *Scandinavian England.* Chichester, pp.131–161.

Wrathmell S. (2012). Northern England: Exploring the Character of Medieval Rural Settlements. In Christie N. and Stamper P., Eds., *Medieval Rural Settlement. Britain and Ireland AD 800–1600.* Oxford, Windgather Press, pp. 249–269.

8

A Viking-Age Site at Workington, Cumbria: Interim Statement

Mike McCarthy and Caroline Paterson

CONTENTS

ABSTRACT   The late Victorian church of St. Michael's, Workington, was largely destroyed by fire in 1994. Before rebuilding started, a programme of archaeological investigation took place. Most of the interior was found to be disturbed by burials but the foundations of the Norman nave and chancel arch were located along with a substantial collection of Anglo-Scandinavian sculpture, some incorporated into the medieval and later fabric. A small number of items of metalwork from the late 9th to the 11th centuries were also found.

The collection of sculpture from this church is one of the largest from Cumbria, suggesting the presence of an important community in the Anglo-Scandinavian period with links to the Hiberno-Norse world around the Irish Sea. In the 12th century, the pre-Norman church was demolished and rebuilt, probably as a simple three-cell parish church to which aisles were added in the later Middle Ages. Further rebuilding work took place in the 18th and 19th centuries.

Introduction

In September 1994, the parish church of St. Michael's, Workington, Cumbria, was very seriously damaged by a fire. The whole of the interior of the Grade II church was gutted, leaving only the tower surviving intact. Immediately after the fire, arrangements were made for the Royal Commission on Historical Monuments to visit the site and make a photographic record prior to any clearance and demolition. Later, the decision to rebuild the church within the surviving shell was made after extensive consultations about its future. Full discussion was held with the Diocesan Advisory Committee who recommended the appropriate actions to the Chancellor, including an archaeological programme of works prior to rebuilding.

The archaeological works began with an evaluation undertaken in June and July 1995, which revealed the foundations of the medieval church and further pieces of sculpture. An excavation followed in

1996–1997 prior to the commencement of the construction programme. All the archaeological work was undertaken under the direction of Paul Flynn for the Carlisle Archaeological Unit with funding provided by the Parochial Church Council and Carlisle City Council. A brief note was published shortly after the work took place (Flynn 1997). The purpose of this chapter is to draw attention to the Viking Age remains at Workington pending a more detailed consideration in the final report.

Background

St. Michael's is the oldest parish church in Workington, a small town on the west coast of Cumbria (Figure 8.1). The church is located on a sand and gravel ridge at a height of 12.4 m Ordnance Datum on the south bank of the River Derwent close to the point where it enters the Irish Sea. To the north and west, the land falls steeply away to the alluvial flats of the Derwent. To the south and east lie an extensive churchyard and the town of Workington. No structural remains of a pre-Norman church have ever been found at Workington, although a number of fragments of sculpture dating to the period were discovered at St. Michael's in the 19th and 20th centuries (Bailey and Cramp 1988). These include pieces dated from the 8th to 9th and 10th to 11th centuries.

Apart from the place-name containing the OE personal name *Weorc* or *Wyrc* (Armstrong et al. 1952, pp. 454–455), the earliest historical record occurs in the *Register of Wetheral*, perhaps from the 1120s. This charter notes the grant of St. Michael's Church by Chetell (or Ketell) to St. Mary's, York (Wilson 1915, pp. 138, 233–234). Chetell is known to have held lands in Allerdale, Copeland, Kendal, and Morland, and was clearly an important figure in Cumbria in the early 12th century, but little is known about him. He is generally regarded as the founder of the Curwen family whose influence and landholdings have been pre-eminent in the area since then.

The tower is the only surviving part of the medieval church of St. Michael's, but the overall layout can be reconstructed from a rough drawing of about 1770 that shows a three-cell structure with a western tower, nave, and chancel. Attached to the south side of the chancel is a chapel with a three-light window. This is probably the Curwen burial place. A small porch is shown attached to the nave. The nave, chancel, burial chapel, and porch were all demolished in 1770 when pressure from an expanding population in Workington led to a rebuilding and enlargement of the church. The new church included a wide nine-bay nave, aisles, a gallery, and a single-bay chancel, as well as the medieval tower, and was completed in 1772. The nave piers rested on the foundations of the medieval nave.

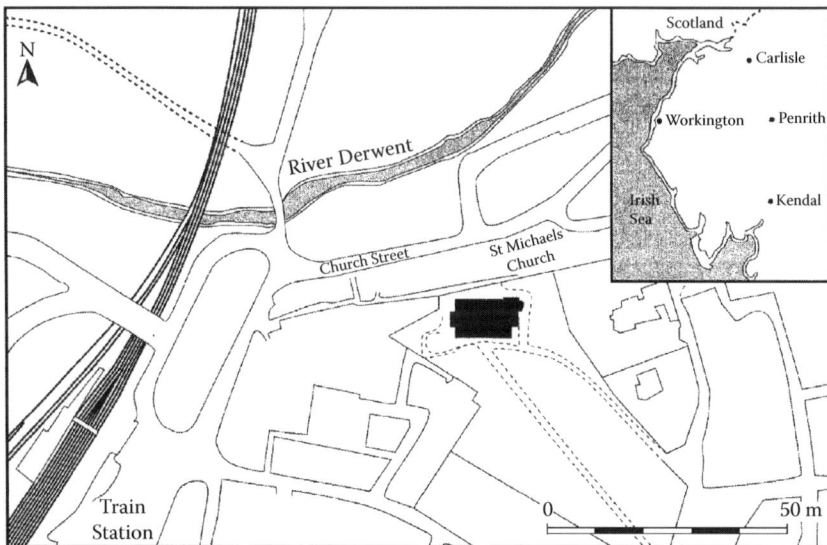

FIGURE 8.1   Location map (drawing by Dan Bashford).

FIGURE 8.2   The nave from the tower showing the medieval nave (white cobbles). Note the pier bases for the 1890 church supporting the metal frame of the gallery are cut into the medieval foundations. Photo: Carlisle Archaeological Unit.

This new church was destroyed by fire in 1887, although once again the medieval tower survived. The church was rebuilt, retaining the medieval tower along with the shell of the 1772 church and was re-consecrated in 1890. In 1994, the church suffered a further disaster when it was burnt down again as described above (Figure 8.2). The medieval tower still survived. Thus, within the space of about two centuries, the principal parish church at Workington has seen four churches on the same site: (1) the medieval church, (2) the late 18th and 19th century church, (3) the Victorian to late 20th century church, and (4) the present church.

Archaeological Results

Phase 1

The archaeological evaluation and subsequent excavation revealed no prehistoric or Roman features, although a small amount of Roman glass including a pillar-moulded bowl was recovered. It is unclear what kind of settlement was associated with the finds, but there is a Roman fort at Burrow Walls, some 1.5 km to the north of Workington, and another at Moresby, about 8 km to the south. Other 'native' sites lie closer to hand but little is known about them.

The earliest substantial feature recognized in the excavation was a ditch 1.1 m wide × 0.72 m in depth, attributed to Phase 1. The ditch is undatable and unattributable to any specific function.

Phase 2

No buildings were identified within the excavation area, but some 35 simple inhumations were found aligned NE to SW, some cut into the backfill of the Phase 1 ditch. A further eight burials had originally been coffined, with fittings composed of iron nails or iron straps located over the chest and ankles. A small number of cist burials were also located as was a 'pillow' burial. Several graves were accompanied by objects of Hiberno-Norse type, including a belt buckle and plate, whetstone, sickle, and knife.

Phase 3

The earliest building identified in the excavation comprised a church with a nave measuring 12 m × 7 m internally and a chancel some 6 m square (Figure 8.2 and Figure 8.3). The east–west alignment of the

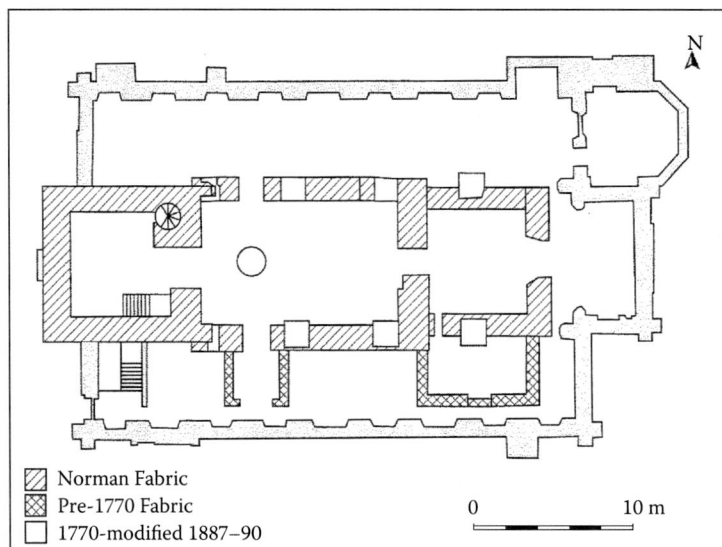

FIGURE 8.3 Plan of St. Michael's Church, Workington showing the Norman, medieval and post-medieval fabric (drawing by Dan Bashford).

church differs slightly from the alignment of the Phase 2 graves. The stratigraphic relationship between nave and chancel was heavily disturbed so that it is not possible to be sure whether they were built at the same time.

The nave and chancel foundations were 2 m wide and made of coursed river cobbles. The opening for the chancel arch was approximately 3 m wide. The chancel was probably square-ended, but its east wall was found to have been substantially destroyed by the construction of the Curwen burial vault before 1770 and then again in 1926 by the building of an underground tunnel to the vault. The demolition of the church in 1770 and subsequent rebuilding were very destructive to the medieval church.

A stratigraphic link between the nave and the tower could not be established, but it is thought that they belong to the same constructional phase. Where the tower foundations below the present ground surface were exposed, they were also seen to be of river cobbles. The east wall of the tower contains a round Romanesque arch with a built-in cushion capital dating it to the 12th century, whilst a double-scalloped capital may have come from the chancel arch (Thurlby 2004, p. 277).

An important discovery was the presence of fragments of cross shafts, hogback grave covers, and other sculptures incorporated in the foundations of the nave, tower, chancel, and porch, making it clear that by the time the church was built, the earlier 10th to 11th century complex had been swept away. A further observation concerns the original entrance to the church. The drawing of about 1770 shows a porch on the south side, but the disposition of graves along the south wall of the nave indicates there was no entrance here, at least initially. Because there was no external access into the tower either, the entrance to the 12th century church must have been on the north side.

A number of graves in Phase 3 can be associated with the early 12th century, including simple inhumations and burials accompanied by stones around the head, sometimes known as 'pillow stones' or 'ear muffs,' as well as stone-lined graves or cists.

Phases 4–7

Additions were made to the tower and a south porch was added, both possibly in the 13th to 14th century. The Curwen burial place was butted up to the chancel in the 14th or 15th century. Around 1450, the fine table tomb of Sir Richard Curwen and his wife Elizabeth was installed in the church, probably in the burial chapel. The church continued to be used for burials into the 19th century.

Grave Types

Some 369 graves were recorded, but there must have been a great many more based on the number of disarticulated remains. The majority were extended inhumations lying supine. Most (244 or 66%) consisted of simple inhumations with no evidence of shrouds, coffins, or grave furniture. These burials seemed to span most periods of use from the pre-Norman to the post-medieval periods.

Twelve burials (3.2%) were in slab-lined graves or cists, some with stone bases and lids, and in one instance the stones were mortared together. Six burials (1.6%) were accompanied either by 'pillow stones' or 'earmuffs' (stones placed to both sides of the head). The remaining 107 (29%) burials showed evidence of shrouds, wooden coffins, and in one case lime, and ranged in date from medieval times to the 19th century. No osteological analysis of the bones has been undertaken at the time of writing.

Sculpture and Metalwork

Some 24 stone sculptural fragments were found during these investigations, making this one of the largest and most important collections of Viking Age sculpture in Cumbria. Detailed comments must await publication of the final report, but here attention is drawn to some salient items.

An almost complete, circular cross-socket decorated with interlace and found in the foundations of the chancel arch is a most unusual discovery (Figures 8.4 and 8.5). An unfinished stone thought to be part of a grave cover rather than a cross was incorporated in the tower foundations. It is decorated with a simple cruciform key pattern (based on the St. Andrew's cross) and two registers of interlace with zoomorphicised Stafford knots, linking it to the 10th century Beckermet group (Figure 8.6).

There are several cross-shaft fragments, including a cross-head with wedge-shaped arms decorated with a lorgnette cross arrangement comprising a central pelleted ring linked to bosses in each arm (Figure 8.7) that has close parallels in a number of places in west Cumbria as well as at Carlisle. A cross-shaft has a vertebral ring chain on two broad faces and a narrow face decorated with two finely incised fronted creatures whose bodies then interlace. The local affinities of this piece include Gosforth and other west Cumbrian sites, but there are also links to the Scandinavian zoomorphic styles. A number of fragments of hogback grave covers were found, including some with animal head terminals, multiple ring chains, and tegulated roof fragments (Paterson 2000).

FIGURE 8.4 Cross-socket embedded in the foundation of the medieval church. Photo: Carlisle Archaeological Unit.

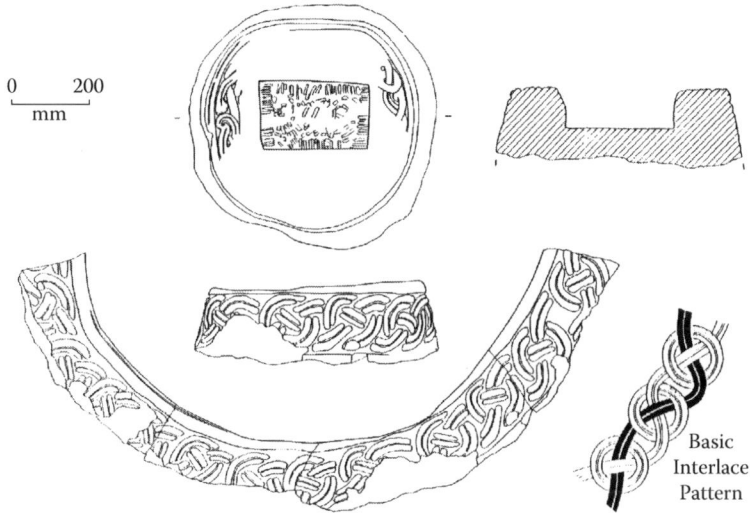

FIGURE 8.5   Decorative detail on the cross-socket (drawing by Philip Cracknell).

FIGURE 8.6   Grave cover in the tower foundations. Photo: Carlisle Archaeological Unit.

The excavation yielded a small number of early medieval artefacts, the contexts of which, although disturbed, reveal that some had accompanied inhumation burials. A gold finger ring is of Viking Age type, with its ends twisted into a knotted terminal and an outer face decorated with punched opposed triangles.

A copper-alloy buckle with a D-shaped loop and a sub-rectangular plate with an axially aligned row of boss-headed rivets belongs to a distinctive series of metalwork fittings from Hiberno-Norse contexts (Figure 8.8a), as does a copper-alloy strap end that still has the remains of a mineralised strap, probably from a waist belt, and which was decorated with simple incised ornament and ring and dot motifs, comparable with examples from the Isle of Man, the Western Isles, and Dublin (Figure 8.8b).

A second strap-end was found with a perforated whetstone and an iron knife. Other utilitarian finds include an iron sickle, clay loom weight and jet spindle whorl. All items can be paralleled in Hiberno-Norse contexts around the Irish Sea including Dublin, the Isle of Man, the Western Isles, and York (Paterson 2001).

The fragments of sculpture reflecting Scandinavian influence indicate that a Christian community was in existence during the 10th century in close proximity to the site of the 12th century church. The

FIGURE 8.7 Cross-head in the foundation of the medieval church. Photo: Carlisle Archaeological Unit.

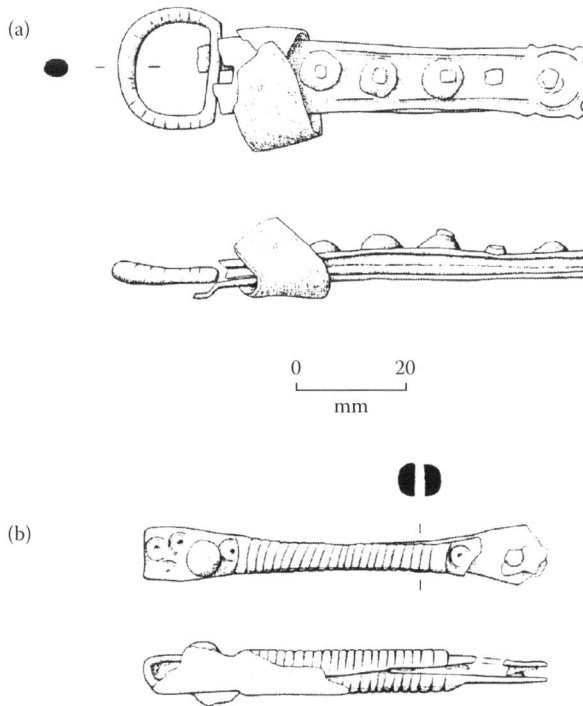

FIGURE 8.8 (a) Copper-alloy D-shaped buckle and plate, (b) copper-alloy strap-end (drawings by Philip Cracknell).

presence of burials with Hiberno-Norse finds indicates that this was a period of transition, with elements of pagan burial practise incorporated within a Christian context. Although contextual details have not been confirmed, it appears that the grave goods were not restricted to elements of clothing or small personal items suspended from clothing fasteners. They included utilitarian objects such as the loom weight and the sickle that reflect the deliberate deposit of grave goods typical of pagan ritual (Gräslund 1987, p. 85).

This material reflection of the transition of pagan Norse settlers to Christian burial practice was observed elsewhere in the region, at St. Patrick's Isle, Peel, Isle of Man (Freke 2001, pp. 73, 96–97) and Carlisle Cathedral (McCarthy et al., this volume; Paterson and Tweddle 2014; D. Tweddle, personal communication). At the latter site, the finds appear to be predominantly restricted to clothing fasteners, perhaps indicating that this community was one step further on in the 'Christianisation' process.

Conclusions

The excavation revealed the plan of the pre-1772 century church at Workington (Figure 8.3). A date for this church is indicated by the presence of sculptural fragments dating to the 10th and 11th centuries in the foundations that provide a crude *terminus post quem* for building in the 12th century or later. However, a Norman date is preferred because of the Romanesque character of the tower arch and the two capitals.

No structural remains were found that could be associated with an 8th to 9th century site or a Viking-age church, although the Phase 1 ditch could be related in some way to an Anglian monastic site. However, as the site has now produced the largest collection of pre-Norman sculpture anywhere in Cumbria (Table 8.1), 24 new fragments out of a total of 172, of which well over 50% is of 10th to 11th century in date, there can be little doubt that a church continued to function here during the period of Scandinavian settlement.

The precise location and form of the early monastic complex and the Viking-age church are matters for speculation, but the sculpture and furnished graves suggest that the 10th century church is very close by. The only clue lies in the alignment of the Phase 2 graves. No other archaeological finds of this period that may narrow down the location are known from Workington, although a reference in the *Historic Environment Record* notes the discovery of a Viking sword at Oysterbanks (West Seaton) on the northern side of the Derwent in 1904.

The location of St. Michael's on the Irish Sea coast next to a major river is consistent with other sites of this date, especially in the north-east. Furthermore, the idea that the body of St. Cuthbert was taken to Workington in its travels around the north, as suggested by Symeon (Arnold 1885, p. 114), may indicate the presence of an ecclesiastical centre if Symeon's comment is accepted.

A short distance upstream along the Derwent is the large parish of Brigham, whose church also yielded a fragment of pre-Viking sculpture. Brigham is thought to have been the ecclesiastical centre of an estate focused on Cockermouth and has produced the second largest collection of 10th to 11th century sculpture in the county (Bailey and Cramp 1988; Winchester 1985, p. 97 and 1987, p. 26). Other estate centres are postulated for Millom and, possibly, St. Bees (Todd 2003; Winchester 2008, pp. 15–17).

The affinities of the sculpture are apparent both in the wider Scandinavian world and locally. Some pieces reflect elements of the Borre, Jellinge, and Mammen styles, but locally much is identical to or has affinities with examples from other places such as Gosforth, Beckermet St. John, and Haile on the coastal strip (Bailey 1980). The quantity of metalwork from St. Michael's may be small, but the D-shaped buckle and its plate (Figure 8.8a) are similar to finds at Carlisle and sites in the western seaways, as are the cist burials and 'pillow stone' or 'ear muff' graves.

Superficially, the quantity, range, and quality of the sculpture as demonstrated in Tables 8.1 and 8.2 combined with a location at the mouth of the Derwent convey the impression of Workington as an

TABLE 8.1

Anglian and Anglo-Scandinavian Sculpture from Cumberland and Westmorland (after Bailey and Cramp 1988 and Paterson 2000)

	6th to 9th Centuries	10th and 11th Centuries	Uncertain Date	Totals
Corpus, Bailey and Cramp 1988[a]	30	114		144
Workington excavations, 1990s	–	24	4	28
Totals	30 (17%)	138 (80%)	4 (3%)	172

[a] Excludes sub-Roman and unprovenanced fragments.

TABLE 8.2

Anglo-Scandinavian Sculpture from Workington, St. Michael's
(after Bailey and Cramp 1988 and Paterson 2000)

Sculpture	Number of Fragments
Cross fragments	11
Grave covers (including hogbacks)	13
Uncertain function	4

important, if not special, place in the 10th and 11th centuries. However, this view can be challenged because the reason we have so much sculpture is an outcome of four major building programmes in 1770–1772, 1887–1894, 1926 (Curwen vault), and 1994. If other churches had undergone so many building programmes, they too might have yielded a good deal more than we have at the moment.

Such a scenario would require us to think of stone sculpture as a standard component of Viking Age churches. Winchester, following Fellows-Jensen, has noted that as many as 15 of the 25 medieval parishes in Copeland produced pre-Conquest sculpture (Winchester 1987, p. 24)—a figure that would certainly increase were the opportunities for discovery present. The idea that the sculpture at Workington was not necessarily special in either quantity or type also finds support by Phythian-Adams, who drew attention to the point that Workington was simply one among a number of miscellaneous estates known as Copeland (ON *kaupaland* = bought land) brought together in the 12th century, forming part of Waltheof's patrimony of Allerdale (Phythian-Adams 1996, pp. 31–33).

Had Workington been an especially significant pre-Conquest estate, we might expect to see this reflected in the presence of high status works in the 12th century, perhaps a castle, but there was none. Quite possibly one of the most dramatic events was the sweeping away of the pre-Norman complex and the construction of a new church in the 12th century. Thereafter, Workington appears to have remained a small fishing settlement on the Cumbrian coast throughout the Middle Ages. It was dominated by a 12th century three-cell church with a tower serving as a baptistery (Thurlby 2004, p. 277).

Whilst aspects of the architectural ornament find reflections at Carlisle, Workington was clearly not in the same league as either Carlisle or Kirby Lonsdale, and it shows little sign of expansion apart from the Curwen burial place. Despite being one of the richest livings in the county (Michael Curwen, personal. communication), there is no evidence that this was reflected in the fabric of the church. Whatever its pre-Conquest promise may have been, Workington appears to have remained small and inconsequential, only acquiring a modicum of prosperity from the mid 17th century with the opening of the local coal mines, a move stimulated in part by the descendants of Chetell, the Curwen family.

ACKNOWLEDGMENTS

The evaluation and the excavation were directed by Paul Flynn for the then Carlisle Archaeological Unit. Juliet Reeves collated much of the archive. The drawings are the work of Philip Cracknell and Dan Bashford. Tim Padley and other members of the Archaeological Unit also contributed in many ways.

We are grateful to Professors Rosemary Cramp and Richard Bailey during and after the excavation for comments on the sculpture and to Dr. Dominic Tweddle for advising on the pre-Norman finds and sculpture. The late Dr. John Todd commented on aspects of the documentary sources, whilst Jenny Jones of the conservation laboratory in the Department of Archaeology at the University of Durham kindly dealt with some of the finds.

Thanks are extended to the Parochial Church Council, especially Canon Terence Sampson, then Rector of St. Michael's Church, Ivy Benn, Linda Hodgson, and John Bailey, the council's architect, whose help and encouragement were critical to the success of this important project. Members of the Cumbria Diocesan Advisory Committee's Working Party on St. Michael's including Mary Burkett and John Packer, the then Archdeacon of West Cumbria, now bishop of Ripon and Leeds, and Michael Curwen provided useful support.

REFERENCES

Armstrong, A.M., Mawer, A., Stenton, F.M. et al. (1952). *The Place-Names of Cumberland I–III*. Cambridge, Cambridge University Press.

Arnold, T., Ed. (1885). *Symeonis Monachi Opera Omnia: Historia Regum. Volume 2,* London.

Bailey, R.N. (1980). *Viking Age Sculpture in Northern England*. London, Collins.

Bailey, R.N. and Cramp, R.J. (1988). *Corpus of Anglo-Saxon Stone Sculpture. Volume II: Cumberland and Westmorland and Lancashire North of the Sands*. London, British Academy.

Flynn, P. (1997). Excavations at St. Michael's, Workington, Church. *Archaeology* 1, 43–45.

Freke, D. (2002). *Excavations on St. Patrick's Isle, Peel, Isle of Man 1982–1988*. Liverpool, Liverpool University Press.

Gräslund, A.S. (1987). Pagan and Christian in the age of conversion. In *Proceedings of 10th Viking Congress, Larkollen, Norway, 1985,* J.E. Knirk, Ed. Oslo, Universitates Oldsaksamling, pp. 81–94.

Paterson, C. (2000). Recent sculptural fragments from St. Michael's Church, Workington. Unpublished report prepared for Carlisle Archaeology.

Paterson, C. (2001). Workington Early Medieval Finds. Unpublished report prepared for Carlisle Archaeology.

Paterson, C. and Tweddle, D. (2014). The 'gold, 'silver,' and copper alloy', In M.R. McCarthy, Ed., A post-Roman sequence at Carlisle Cathedral. *Archaeology Journal* 170 (in press).

Paterson, C., Parsons, A.J., Newman, R.M., Johnson, N. and Howard-Davis, C. (2014). *Shadows in the Sand: Excavation of a Viking-Age Cemetery at Cumwhitton, Cumbria* (in press).

Phythian-Adams, C. (1996). *Land of the Cumbrians: A Study in British Provincial Origins AD 400–1120*. London, Scholar Press.

Thurlby, M. (2004). Romanesque architecture and architectural sculpture in the Diocese of Carlisle. In *Carlisle and Cumbria: Roman and Medieval Architecture, Art, and Archaeology. British Archaeological Association Conference Transactions* XXVII, pp. 269–290.

Todd, J. (2003). The pre-Conquest church in St Bees: A possible minster? *Transactions of the Cumberland Westmorland Antiquarian Archaeological Society* 3, 97–108.

Wilson, J., Ed. (1915). *The Register of the Priory of St. Bees*. Publications Surtees Society, Vol. 126.

Winchester, A.J.L. (1985). The multiple estate: a framework for the evolution of settlement in Anglo-Saxon and Scandinavian Cumbria. In *The Scandinavians in Cumbria,* J.R. Baldwin and I.D. Whyte, Eds. Edinburgh, Scottish School for Northern Studies, pp. 89–101.

Winchester, A.J.L. (1987). *Landscape and Society in Medieval Cumbria*. Edinburgh, John Donald.

Winchester, A.J.L. (2008). Early estate structures in Cumbria and Lancashire. *Medieval Settlement Research Group* 23, 14–21.

9

Were There Vikings in Carlisle?

Mike McCarthy, Janet Montgomery, Ceilidh Lerwick, and Jo Buckberry

CONTENTS

ABSTRACT   Excavations at Carlisle Cathedral revealed a sequence of activities spanning the 4th to 12th centuries AD. The finds included a long-lived cemetery under the former nave comprising inhumations aligned east–west. Unfortunately, the cemetery was heavily disturbed in medieval and post-medieval times but coins, metalwork, and other items along with radiocarbon determinations established that it was in use in the Viking Age.

Isotopic analyses indicated that some of the individuals enjoyed terrestrial diets. The cemetery, artefacts, and complete archaeological sequence are of great importance to Carlisle because they show it to be an important town between the late 9th and 12th centuries and thus shed light on its pre-urban origins. Although it is not possible to identify individuals of Scandinavian origin, the results are significant additions to our understanding of settlement in Cumbria.

Introduction

A Viking presence in Cumbria has long been acknowledged thanks to the work of W.G. Collingwood and the Reverend W.S. Calverley in the late 19th and early 20th centuries (Griffiths and Harding, Chapter 1). Their studies of the many examples of sculpture in the county have been followed up, systematized, and thoroughly assessed by Bailey and Cramp (Bailey 1980 and 1985; Bailey and Cramp 1988).

Cumbrian place-names with Scandinavian-derived elements such as *-bý, -sætr, -skáli,* and *-þveit* have also been the subjects of many studies, most recently by Gillian Fellows-Jensen who subjected the distribution of Scandinavian names in north-west England, Dumfries, and Galloway to critical scrutiny (Fellows-Jensen 1985a and b; Jesch, Chapter 3). From time to time, Viking Age graves or artefacts have been found at Hesket, Flusco near Penrith, Aspatria, and Cumwhitton (Cowen 1934, 1948, and 1967; Edwards 1998; Brennand 2006; Paterson et al. 2013).

Other studies have sought to place Scandinavian settlements in north-west England in their wider historical and geographic contexts, especially the Irish Sea region (Griffiths 2004; Higham 2004; Newman 2006). To date, relatively little fieldwork has been done on Viking Age sites, and the position of Carlisle in the period of Scandinavian settlement has largely escaped attention. The aim of this chapter is to consider this issue with a focus on two questions. First, why would the Vikings be interested in Carlisle's status and location? The second question is whether we can recognize skeletal remains as of Scandinavian origin via the analysis of isotopes derived from bones and teeth.

Carlisle's Status and Location

It is widely believed that one of the main motivations for Viking attacks on Britain and Ireland from the late 8th century was the loot to be gained from raiding monasteries, some of which were rich and poorly defended from the sea, and thus represented relatively easy targets. However, whilst the impact of the raids were considerable, it is far from clear that they were always as destructive as some commentators have alleged.

It is difficult to make an assessment of impacts on north-west England because of the dearth of early historical sources for the western parts of the Kingdom of Northumbria. By the late 8th century, a number of Northumbrian monastic communities were well established west of the Pennines in Carlisle, Dacre, Cartmel, Brigham, Workington, and Heysham, and Hoddom and Whithorn north of the Solway Firth. Several of these may have presented attractive targets for Viking raiders.

One of the best documented monastic sites is at the former Roman town of Carlisle (Figure 9.1). The two *vitae* of St. Cuthbert tell us of its existence by AD 685, the occasion of a visit by St. Cuthbert with Iurminburh, the Northumbrian queen (Colgrave 1969). Carlisle and the other monastic communities must have resembled small nucleated settlements, the centres of estates with the religious at their hearts.

At Carlisle a reeve (*praepositus*) named Waga oversaw the management of the estate and probably the lay population. Carlisle was a double monastery (a minimum of two churches, one for each sex). We still have little knowledge about other communities west of the Pennines. Some excavations have taken place such as at Dacre (Leech and Newman 1985), but very little formal investigation has been conducted elsewhere.

One uncertainty about all these sites is whether they survived the period of Viking raiding and subsequent settlement. Abbot Eadred, whose *floruit* was apparently in the 850s to the 880s period that spanned

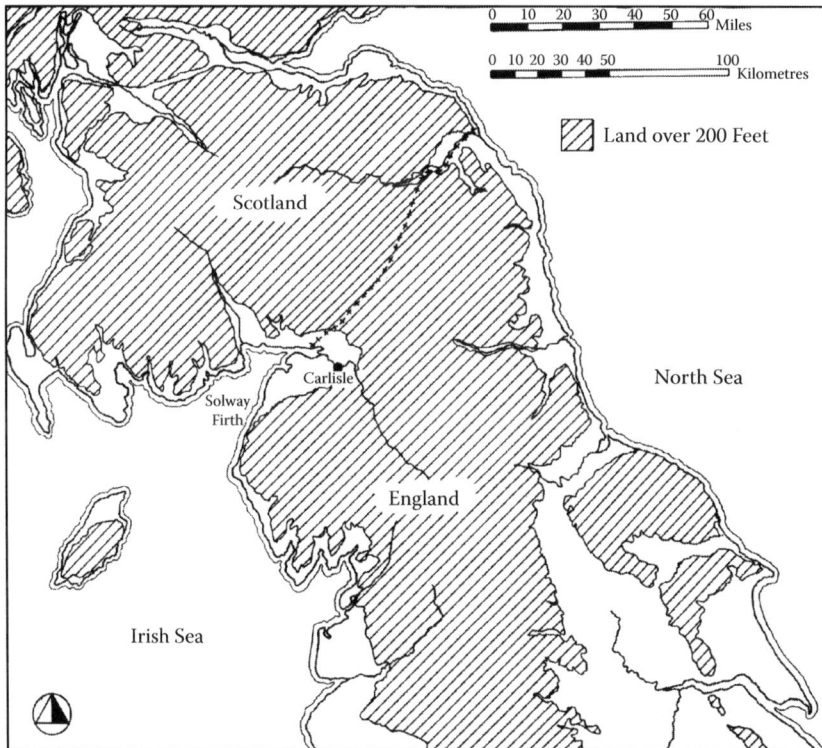

FIGURE 9.1   Location of Carlisle. (Drawing courtesy of Philip Cracknell.)

the Viking take-over of Northumbria, was said by Symeon to be from Carlisle (Arnold 1885, p. 114). The Eadred name also appears several times in the Durham *Liber Vitae*. Eadred, an abbot of Carlisle, was also said to have given estates to St. Cuthbert at Chester-le-Street in about 883 (Hart 1975, p. 156), although how or why a Cumbrian cleric should distribute lands in the north-east is difficult to explain.

Even so, notwithstanding Symeon's partisan attitudes about the Cumbrian estates formerly held by Lindisfarne and of particular interest to his employer at Durham (Rollason 1998; Sharpe 1998), the presence of an abbot named Eadred in mid 9th century Carlisle is certainly not implausible. Acceptance of that fact extends the life of the Carlisle double monastery into the second half of the 9th century.

We have few details of monastic communities in the west. Some may have been small and relatively impoverished, The case for believing that Carlisle was better endowed than most is based on the idea that a monastic community was established in a formerly important Roman town, *Luguvalium*, and that it was chosen as the setting for Queen Iurminburh's sojourn as she awaited the outcome of the battle of her husband, Ecgfrith against the Picts. What is interesting here is not just that we have an unusual body of anecdotal information about Carlisle in the *vitae*, but that it was selected as the venue for Iurminburh's wait, perhaps because she came from there.

If that was the case, we can surmise that Carlisle was the home of an important family with significant royal connections. We know little about Iurminburh but Stephen's comments in his *Life of Wilfrid* certainly suggest that she was a formidable lady (Colgrave 1985) and was remembered in later times as the appearance of her name in the 9th century Durham *Liber Vitae* suggests. These are small points but they lend weight to the idea that Carlisle, whilst not necessarily as wealthy as Ripon or Hexham, nevertheless occupied a relatively prominent position amongst western Northumbrian monasteries.

With this in mind, and given the Vikings' much written-about predilection for loot, it seems inherently unlikely that Carlisle would have been bypassed or left unscathed. The one source that might provide a clue is John of Worcester, writing in the early 12th century, who tells us that the Danes destroyed Carlisle which then lay deserted for almost 200 years (McGurk 1998, pp. 62–63). The first part of John's statement concerning Danish destruction, although not independently attested, may well be correct even though we cannot tell when it occurred. Smyth's analysis of the movements of Viking kings and their armies draws attention to the Danish king, Ivarr the Boneless, who went from the Midlands to Dumbarton in 870—a route that may have taken him by way of Stainmore and the Eden valley, and by implication through Carlisle (Smyth 1977, p. 258). Ivarr was not a man to pass up an opportunity!

Carlisle's interest to the Vikings is unlikely to have been confined to loot. An important factor was its location at the northern end of the Eden valley, close to the silt-choked Solway Firth and the west coast (McCarthy 2010, pp. 105–109). It stood astride the only major landward route in the west between what is now England and Scotland. On the English side, the fertile Eden valley was wedged between poor glacial tills, mosses, and uplands. On the Scottish side, routes go north by way of Annandale or Eskdale, also fringed with relatively poor soils.

In the middle of this is Carlisle and he who holds that, to coin a phrase, is in command of that crucial, landward north–south route. Its importance was earlier recognized by the Romans, and later by the earls of Northumbria, the kings of England, and the Scots (McCarthy 2010, pp. 109–118). From the later 7th century, it was the centre of an important royal estate. After the late 9th century it served as a crossroads or nodal point between the Danes in Yorkshire and eastern Northumbria, the Strathclyde Norse centred on Govan and the Clyde valley, and the Irish Norse sailing from Dublin following their expulsion in AD 902. Amongst the refugees fleeing to eastern Northumbria from west of the Pennines, we are told by the anonymous *Historia de Sancto Cuthberto*, were Abbot Tilred of Heversham (Cumbria), Alfred son of Brihtwulf, and possibly also Eadred son of Ricige, who received lands in Durham from Bishop Cutheard of Lindisfarne (Craster 1954, p. 190; Hart 1975, pp. 162–163; Whitelock 1979, p. 287).

Until recently knowledge of Viking Age Carlisle was very limited. However, after excavations at the cathedral in 1988, the situation becomes a little clearer (McCarthy 2014). Figure 9.2 shows the west end of the cathedral. Figure 9.3 shows excavations in progress. The Dean and Chapter commissioned the excavation in advance of building a new underground treasury at the west end.

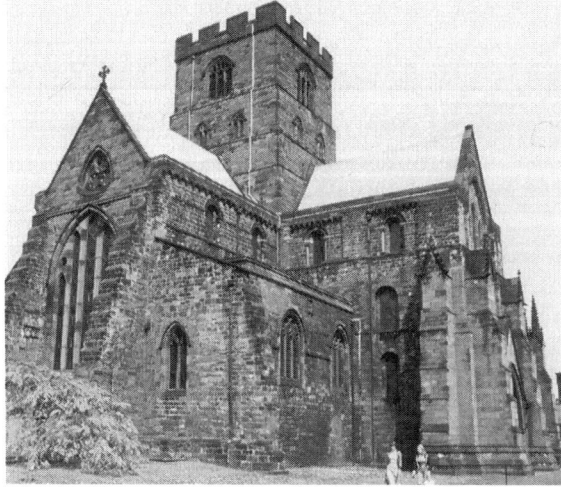

FIGURE 9.2   West end of Carlisle Cathedral. The 12th century structure was built over a cemetery containing graves of Anglo-Scandinavian date. (Photograph courtesy of Mike McCarthy.)

FIGURE 9.3   Excavations in progress in 1988. (Photograph courtesy of Graham Keevill.)

Excavation revealed six phases of archaeology from Roman to medieval periods. The monastic site associated with St. Cuthbert would have been a clear target, but the excavations have not revealed anything structural with which it can be linked. Geophysical survey located a substantial anomaly below the former 19th century St. Mary's Church that is thought to have been part of the complex, but confirmation of this is needed.

A pre-Norman cemetery of 41 graves was discovered and may include Anglian burials, but the radio-carbon dates available to date are ambiguous on that point (Batt 2014; McCarthy 2014). Whilst the overall quantity of pre-Anglo Scandinavian finds is limited, a third or more of the *stycas* known from Carlisle were recovered during the cathedral excavations. It is conceivable, therefore, that if the monastic site lay at the eastern end of the cathedral precinct, the western or north-western end as open ground may have functioned as a market. Metcalf long ago made a similar point about the incidence of *stycas* on ecclesiastical sites (Metcalf 1987).

The second part of John of Worcester's statement about Carlisle's desertion for 200 years after a Danish raid can now be seen as manifestly untrue. The cemetery excavation revealed furnished buri-als of the late 9th to early 10th centuries and clearly demonstrates the presence of a community at that time. Finds included gold, silver, and copper alloy objects, ironwork, and an antler comb with strong Hiberno-Norse affinities (Paterson and Tweddle 2103). The objects were personal or domestic in char-acter, reflecting high or relatively high status and contrast with the warrior-like weapon burials located at Cumwhitton and found elsewhere in Cumbria as at Hesket, for example (Paterson et al. 2013). In that respect, the cathedral burials are closer to those at St. Michael's Workington (McCarthy and Paterson, Chapter 8).

The ethnic nature of that community is of some interest, and to that end an examination of skeletal material combined with an analysis of isotopic data was undertaken with a view to seeing whether any individuals could have originated outside the immediate area of Carlisle. The osteological and isotopic data, limited in size though it is, includes men, women, and children, but revealed no sign of individuals brought up in a maritime environment as might be expected of Scandinavian immigrants. The sample size available, however, was very small and we should not attach too much weight to nega-tive evidence. Even so, because Carlisle is at a nodal point, a population of mixed ethnicity would not be unexpected.

Osteology

A number of burials can be assigned to the 9th to 11th centuries stratigraphically. Some were furnished with metalwork, mostly belt fittings (Paterson and Tweddle 2013), and 8th to 11th century radiocarbon dates have been obtained from teeth dentine and ribs (Batt 2014). The excavated population was quite small, and spans approximately 1400 years, so making meaningful inferences about the population is difficult.

It consisted of children, infants, and male and female adults (Lerwick and Buckberry 2013). Females outnumbered males in the sample analysed, but the sample excavated and analysed may not have been representative of the buried population. The pre-Norman sample of burials includes male and female adults, many aged at least 45 years. No examples of trauma or other significant pathologies were pres-ent, but rather the population suffered more common conditions including dental caries, calculus, joint disease, and maxillary sinusitis.

Cranial form is influenced by both genetics and environment (Mays 2000). The cranial indices for Carlisle vary considerably, suggesting the inhabitants were biologically diverse. Notably, the average values at Carlisle, indicating mesocrany (medium-headed), differ from those recorded for the Anglo-Scandinavian populations at Riccall Landing and Masham, north Yorkshire that contained largely doli-chocranic (long) skulls (Lerwick and Buckberry 2013; 2014).

However, cranial form is known to vary over time. Dolichocranic skulls are common in early medi-eval and post-medieval populations and brachycranic (round) and mesocranic skulls are the norms for medieval populations. Thus the variation in cranial indices seen at Carlisle may be explained by the wide variation in dates for the few skeletons with intact crania excavated and analysed at the site. It is too early to assess the significance of these observations because the number of individuals is too small and not all have been dated securely.

Isotopes

The ribs of 18 skeletons were subjected to carbon and nitrogen isotope analysis of bone collagen to gain information about dietary protein consumption rather than whole diet data (Ambrose and Norr 1993; Jim et al. 2004). Teeth were also subjected to carbon and nitrogen isotope analysis of dentine collagen and strontium, oxygen, and carbon isotope analysis of enamel. Carbon and nitrogen isotope analysis is used routinely to infer diet rather than origin. However, if a diet consumed in an individual's place of origin was significantly different from that at the place of burial due to differing access to certain types of foods or cultural norms, diet may be used at times to identify migrants.

For example, before the 20th century, inhabitants of Britain and Ireland had extremely limited or no access to C_4 plants such as maize, millet, sugar cane, and sorghum. Consequently, when bone materials found at archaeological sites reveal carbon isotope ratios indicative of C_4 diets, such as those found in Roman York (Müldner et al. 2011) and 19th century Ireland (Beaumont et al. 2013), the most plausible conclusions are that the individuals ate certain foods elsewhere or their food was imported in significant quantities.

In these two examples, the Roman individuals were deemed to have migrated to England from the Continent (Müldner et al. 2011). The Irish individuals were thought to have consumed maize in the form of famine relief imported from the United States during the Irish Potato Famine (Beaumont et al. 2013).

In the Viking period, it is possible that immigrants to Britain from Scandinavia may be identified because they joined a population that showed no significant consumption of marine protein, even in coastal regions, until the 14th century (Barrett et al. 2001; Jay and Richards 2007; Müldner and Richards 2007). For example, stable isotope ratios indicative of marine protein consumption were obtained from two male Viking burials at Westness (Barrett and Richards 2004). Oxygen isotope analysis of their tooth enamel strongly suggested origins in a colder, more northerly location than the British Isles (Montgomery et al. 2014).

At Carlisle, it was hypothesised that if individuals were found amongst the Anglo-Scandinavian population with evidence for the consumption of marine protein during childhood (dentine) or in later life (adulthood), it would be a reasonable conclusion that these individuals were of Scandinavian origin, although the absence of marine protein does not in itself preclude its consumption. An indication of when migration occurred could also be obtained by comparing the results of dentine and bone analysis for each individual.

As would be expected in England before the 14th century, most skeletal samples at Carlisle are strongly suggestive of a diet composed of solely or predominantly of C_3 terrestrial protein (Jay and Richards 2007; Müldner and Richards 2007). However, the data (Figure 9.4) also show increases in $\delta^{15}N$ and $\delta^{13}C$ over time for the human skeletal material, but all the individuals with $\delta^{13}C$ values higher than 20.2‰ may be attributed to the late medieval and modern periods (Montgomery and Towers 2014).

This highlights the importance of establishing date and context when attempting to use this type of analysis to infer origins. Only a small number of skeletons have been dated, but the trend at Carlisle is consistent with other datasets from Britain, and is probably associated with the onset of increased fish consumption in Britain during the 14th century (Müldner and Richards 2007).

Carbon and nitrogen isotopes for human bone and dentine results are found across the whole human range. For three of the individuals, the difference between childhood (dentine) and adult (bone) values is negligible, indicating no measurable change in dietary protein that might have been ascribed to different origins. The remaining two individuals exhibited a difference in $\delta^{15}N$ of *ca.*1‰ between dentine and bone collagen, although the shift is in opposite directions. This suggests the shifts indicate different dietary changes and the changes cannot both be ascribed to the same process. Neither result, however, would indicate a significant change from a childhood diet rich in marine fish to one based on terrestrial protein.

Strontium and oxygen isotope data for investigating origins were obtained from the enamel (Montgomery and Towers 2014). Unfortunately, only five of the skeletons from Carlisle had extant teeth and only one of them one dated to the Anglo-Scandinavian period. These two isotope systems are useful

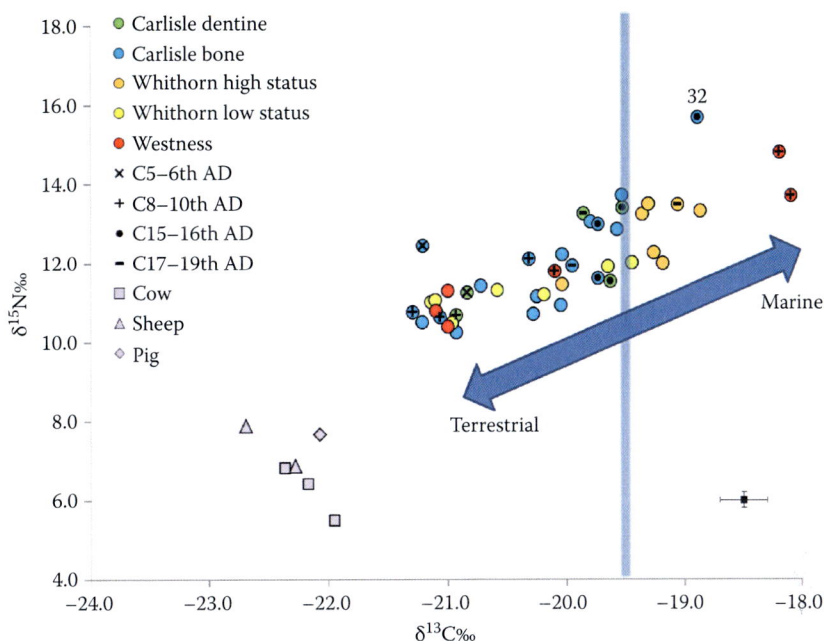

FIGURE 9.4 Carbon and nitrogen isotope data for humans and animals found at Carlisle Cathedral excavation. Individuals to left of line show δ¹³C values indicating their protein came from predominantly terrestrial sources.

to provide information on the geological terrains from which foods, particularly plant foods, are sourced (strontium) and the climate in the place of origin (oxygen).

For immigrants to England from Norway and Sweden, it is reasonable to assume that both strontium and oxygen may identify them because the geology of Norway and Sweden is very different. The predominant ancient granitic rocks of the Baltic Shield that should produce high strontium ratios are not found commonly in England (Evans et al. 2010 and 2012). The oxygen isotope ratios of their drinking water should reflect the colder winters and more northerly location of Scandinavia (Fricke et al. 1995).

Virtually all the strontium isotope ratios obtained to date for humans in Britain are below 0.7150 (Evans et al. 2012). Enamel values above 0.7150 require the consumption of plants sourced from soils derived from granitic rocks such as those found in Norway and Sweden. Given the current weight of evidence, it is therefore reasonable to assume that individuals in the Viking period with such high values were more likely to be of Scandinavian than British origin.

However, there are considerable areas of overlap in both strontium and oxygen isotope ranges between Britain and Ireland and between these two places and Scandinavia (Montgomery et al. 2014). Biosphere strontium isotope ratios can be affected by a range of factors such as rainfall and proximity to a coast. As argued on the literature, it is possible that individuals originating in coastal or high rainfall areas of Norway may not be distinguishable from individuals originating in similar environments in Britain and Ireland based on strontium isotope evidence alone. This observation will be clarified after the results from burials in the Scandinavian homelands undertaken for the 'Viking Migration in the North Atlantic' project directed by T.D. Price are published.

It is very rare indeed to obtain δ¹⁸O$_p$ values below 16.5‰ for humans excavated in Britain or Ireland where higher values between 16.5 and 18.5‰ are usually expected (Evans et al. 2012). The expected difference is even greater between Scandinavia and the western seaboard of Britain where δ¹⁸O$_p$ values are highest. Nonetheless, values below 16.0‰ have been obtained from male Viking-type burials in Dublin and Orkney (Montgomery et al. 2014) and from two apparently catastrophic male burial assemblages assumed to be Viking raiding parties from Oxford (Pollard et al. 2012) and Weymouth (Chenery et al. 2014).

For a few of these individuals, both strontium and oxygen isotope ratios were highly unusual for Britain or Ireland, but for most, the range of strontium isotopes obtained was consistent with many areas of Britain and Ireland, suggesting that using strontium alone is not sufficient in most cases to discriminate indigenous Briton from immigrant Scandinavian. A similar range of strontium isotope ratios was obtained for the early populations of Iceland, where indigenous individuals should have revealed a low (<0.7090) and relatively unique range of strontium isotope ratios derived from the basaltic geology (Price and Gestdottir 2006). For many of these Icelandic burials, therefore, it is difficult to know whether the individuals were born in Scandinavia, Ireland, or Britain based on strontium alone.

For the Carlisle skeletons, the strontium isotope data parameters ranged from 0.7096 to 0.7106 and the concentrations from 51 to 113 ppm (Montgomery and Towers 2014). Both are within the normal range of values that would be expected for people inhabiting a region of sedimentary Triassic rocks such as those found in the Carlisle region on the western seaboard of Britain (Evans et al. 2012). Such isotope ratios are, however, very common and can be found in many locations across Britain, Ireland, and further afield (Montgomery et al. 2014). They are thus not unique in any way to the Carlisle area. For the skeleton dating to the Anglo-Scandinavian period, no evidence from oxygen isotopes indicated origin outside Britain (Montgomery and Towers 2014).

Discussion

The archaeology of the cathedral unequivocally demonstrates the existence of a community at Carlisle prior to the foundation of the Augustinian priory and the diocese in the early 12th century. Coins of Athelstan and Aethelred II, together with metalwork including copper alloy belt fittings (Figure 9.5), a gold toggle, and probable iron-bound coffins all indicate a date for this community in the Anglo-Scandinavian period, confirmed by a small number of radiocarbon determinations (Batt 2014). Furthermore, subsequent geophysical work provides hints that the cemetery was extensive, and probably belonged to a church thought to lie at the eastern end of the Cathedral precinct.

Neither isotopic nor osteological analyses provided evidence of non-local origins, although the possibility of some biological diversity is raised. There is no suggestion, for example, of a marine protein diet as was found in two burials excavated from a Viking cemetery at Westness in the Orkney Islands. At Whithorn, Galloway, marine protein was part of the diets of high status 14th to 15th century clerics, but the lay population even at a location so close to the sea, enjoyed a terrestrial diet, as was also the case at Carlisle (Montgomery et al. 2009).

It is currently difficult to interpret these data beyond suggesting two possibilities. The first is that the Viking Age population of Carlisle was largely of local origin, and the second suggests a hybrid population in which locals lived side by side with incomers from Ireland, Scandinavia, Scotland, or the Danelaw. This latter possibility is in accord with isotopic studies of skeletal data from Dublin, Westness, and Trelleborg that strongly imply that Viking communities, and armies in particular, were not necessarily homogeneous groups, but were of diverse origins (Montgomery et al. 2014; Price et al. 2011).

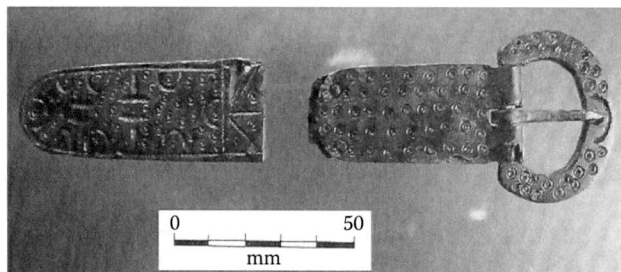

FIGURE 9.5 Copper alloy belt buckle and strap end of early 10th century date. Similar belt suites are known from York and the Isle of Man. (Photograph courtesy of Philip Cracknell.)

Finally, there is the question posed in the title of this paper: Were there Vikings in Carlisle? At present, the answer is equivocal. On the one hand, a population of high status individuals of mixed ethnicity would not be surprising based on Carlisle's location. The found metalwork certainly has affinities both in the Danelaw and in the Western Isles. The discovery of an oval brooch and other items of Hiberno-Norse type at Cumwhitton, only 3 miles from Carlisle, also supports the idea that Vikings were present nearby (Griffiths, Chapter 2; Lee, Chapter 4).

Viking Age sculpture is known from Carlisle and Stanwix (Bailey and Cramp 1988, pp. 84–87 and 147). Place-names also indicate a strong Scandinavian presence in Cumbria, but they too are mostly distributed around the west coast, Lakeland, and the upper Eden valley rather than in the immediate vicinity of Carlisle (Fellows-Jensen 1985). On the other hand, neither osteological nor isotopic data have helped us identify non-local individuals with certainty. Finally, we may note that three stones forming part of the 12th century fabric of the cathedral bear runes in Old Norse. As there would be little point in creating an inscription that could not be read, we may conclude that Old Norse could be understood in the local community (McCarthy 2014).

If the pre-Norman cemetery below the cathedral is indeed extensive, it is likely to have potential for future research on the ethnic mix of the Carlisle population. For this to be realized, it will be necessary to identify areas that have witnessed less post-Norman disturbance. This is not an insignificant task because the Anglo-Scandinavian period lay at the interface between the kingdoms of the Scots and the English and occupied a time when the seeds of urbanization were again taking root.

ACKNOWLEDGMENTS

The osteological isotopic analysis and radiocarbon dating of human remains from the cathedral excavations in 1988 were undertaken with the help of a research grant provided by the British Academy (SG 100732). The map (Figure 9.1) and photograph of the belt suite (Figure 9.4) were prepared by Philip Cracknell. Figure 9.2 is by Mike McCarthy and Figure 9.3 is by Graham Keevill who directed the excavations. Figure 9.5 was prepared by Janet Montgomery.

REFERENCES

Ambrose, S. and Norr, L. (1993). Experimental evidence for the relationship of the carbon isotope ratios of whole diet and dietary protein to those of bone collagen and carbonate. In *Prehistoric Human Bone, Archaeology at the Molecular Level*, Lambert, J.B. and Grupe, J., Eds. Berlin, Springer, pp. 1–37.

Arnold, T., Ed. (1882). *Symeonis Monachi Opera Omnia, Historia Regum. Volume 1*. London.

Arnold, T., Ed. (1885). *Symeonis Monachi Opera Omnia, Historia Regum. Volume 2*. London.

Bailey, R.N. (1980). *Viking Age Sculpture*. London, Collins.

Bailey, R.N. (1985). Aspects of Viking Age sculpture in Cumbria. In *The Scandinavians in Cumbria*, Baldwin, J.R. and Whyte, I.D., Eds. Edinburgh, School of Scottish Studies, pp. 53–64.

Bailey, R.N. and Cramp, R.J. (1988). *The British Academy Corpus of Anglo-Saxon Sculpture, II. Cumberland, Westmorland and Lancashire North-of-the-Sands*. Oxford, Oxford University Press.

Barrett, J.H., Beukens R.P., and Nicholson, R.A. (2001). Diet and ethnicity during the Viking colonization of northern Scotland: evidence from fish bones and stable carbon isotopes. *Antiquity*, 75, 145–154.

Barrett, J.H. and Richards, M.P. (2004). Identity, gender, religion, and economy: new isotope and radiocarbon evidence for marine resource intensification in early historic Orkney, Scotland. *European Journal of Archaeology*, 7, 249–271.

Batt, C.M. (2014). Radiocarbon dates. In *A Post-Roman Sequence at Carlisle Cathedral,* McCarthy, M.R., Ed. *Archaeological Journal*, 171, pp. 187–259.

Beaumont, J., Geber, J., Powers, N. et al. (2013). Victims and survivors: stable isotopes used to identify migrants from the Great Irish Famine to 19th century London. *American Journal Physical Anthropology,* 150, 87–98.

Brennand, M. (2006). Finding the Viking dead. *Current Archaeology,* 204, 623–629.

Chenery, C., Evans, J.A., Score, D. et al. (2014). A boat load of Vikings? *Journal of the North Atlantic* (in press).

Colgrave, B. (1969). *Two Lives of St. Cuthbert: Text Translation and Notes.* New York, Greenwood Press and Cambridge, Cambridge University Press.

Colgrave, B. (1985). *The Life of Bishop Wilfrid by Eddius Stephanus: Text Translation and Notes.* Cambridge, Cambridge University Press.

Cowen, J.D. (1934). A catalogue of objects of the Viking period in the Tullie House Museum, Carlisle. *Trans-Cumberland Westmorland Antiquarian Archaeological Society,* 2, 166–187.

Cowen, J.D. (1948). Viking burials in Cumbria. *Trans-Cumberland Westmorland Antiquarian Archaeological Society,* 48, 73–76.

Cowen, J.D. (1967). Viking burials in Cumbria: a supplement. *Trans-Cumberland Westmorland Antiquarian Archaeological Society,* 67, 31–34.

Craster, E. (1954). The patrimony of St. Cuthbert. *English Historical Review* 271, 177–199.

Edwards, B.J.N. (1998). *Vikings in North-West England: The Artifacts.* Lancaster, Centre for North-West Regional Studies.

Evans, J., Montgomery, J., Wildman, G. et al. (2010). Spatial variations in biosphere $^{87}Sr/^{86}Sr$ in Britain. *Journal of the Geological Society,* 167, 1–4.

Evans, J., Chenery, C.A., and Montgomery, J. (2012). A summary of strontium and oxygen isotope variation in human tooth enamel excavated from Britain. *Journal of Analytical Atomic Spectroscopy,* 27, 754–764.

Fellows-Jensen, G. (1985a). *Scandinavian Settlement Names in the North-West.* Copenhagen, C.A. Reitzels Forlag.

Fellows-Jensen, G. (1985b). Scandinavian settlement in Cumbria and Dumfriesshire: place-name evidence. In *The Scandinavians in Cumbria,* Baldwin, J.R. and Whyte, I.D., Eds. Edinburgh, School of Scottish Studies, pp. 65–82.

Fricke, H.C., O' Neil J.R., and Lynnerup, N. (1995). Oxygen isotope composition of human tooth enamel from medieval Greenland: linking climate and society. *Geology,* 23, 869–872.

Griffiths, D. (2004). Settlement and acculturation in the Irish Sea Region. In *Land, Sea, and Home,* Hines, J., Lane, A., and Redknap, M., Eds. Society of Medieval Archaeology Monograph 20, pp. 125–138.

Hart, C.R. (1975). *The Early Charters of Northern England and the North Midlands.* Leicester, Leicester University Press.

Higham, N.J. (2004). Viking Age settlement in the north-western countryside: lifting the veil. In *Land, Sea, and Home,* Hines, J., Lane, A., and Redknap, M., Eds. Society of Medieval Archaeology Monograph 20, pp. 297–311.

Jay, M. and Richards, M.P. (2007). British Iron Age diet: stable isotopes and other evidence. *Proceedings of Prehistoric Society,* 73, 169–190.

Jim, S., Ambrose, S.H., and Evershed, R.P. (2004). Stable carbon isotopic evidence for differences in the biosynthetic origin of bone cholesterol, collagen, and apatite: implications for their use in palaeodietary reconstruction. *Geochimica et Cosmochimica Acta,* 68, 61–74.

Leech, R. and Newman, R. (1985). Excavations at Dacre, 1982–1984: interim report. *Transactions of the Cumberland and Westmorland Antiquarian and Archaeology Society,* 85, 87–93.

Lerwick, C. and Buckberry, J. 2013. The human remains at Carlisle Cathedral. In Ed. M.R. McCarthy Excavations at Carlisle Cathedral in 1988: Roman, Medieval and Post Medieval Data. [data set] York Archaeology Data Service [distributor] doi: 10.5284/1019911.

Lerwick, C. and Buckberry, J. (2014). The human remains. In *A post-Roman Sequence at Carlisle Cathedral,* McCarthy, M.R., Ed. *Archaeological Journal,* 171, 187–259.

Mays, S. (2000). Biodistance studies using craniometric variation in British archaeological skeletal material. In *Human Osteology in Archaeology and Forensic Science,* Cox, M. and Mays, S., Eds. London, Greenwich Medical Media, pp. 277–288.

McCarthy, M.R. (2010). Carlisle: the making of a medieval town on the Anglo-Scottish frontier. In *Making A Medieval Town: Patterns of Early Medieval Urbanization,* Buko, A. and McCarthy, M.R., Eds. Warsaw, Institute of Archaeology and Polish Academy of Sciences, pp. 105–129.

McCarthy, M.R. (2014). A post-Roman sequence at Carlisle Cathedral. *Archaeological Journal,* 171 187–259.

McGurk, P. (1998). *The Chronicle of John of Worcester, Volume III: The Annals for 1067–1140.* Oxford, Clarendon Press.

Metcalf, D.M. (1987). A topographical commentary on the coin finds from 9th century Northumbria (ca. 780–870). In *Coinage in Ninth-Century Northumbria.* Tenth Oxford Symposium on Coinage and Monetary History, Metcalf, D.M. Ed. Oxford, British Archaeological Series 180, pp. 361–382.

Montgomery, J., Muldner, G., Cook et al. (2009). Isotope analysis of bone collagen and tooth enamel. In *Clothing for the Soul Divine: Burials at the Tomb of St. Ninian. Excavations at Whithorn Priory, 1957–1967*, Lowe, C., Ed. Edinburgh, Historic Scotland, pp. 63–80.

Montgomery, J., Grimes, V., Buckberry, J. et al. (2014). Finding Vikings with isotope analysis: the view from wet and windy islands. *Journal of the North Atlantic* (in press).

Montgomery, J. and Towers, J. (2014). The isotopes. In *A Post-Roman Sequence at Carlisle Cathedral*, McCarthy, M., Ed. *Archaeological Journal*, 171, 187–259.

Müldner, G. and Richards, M.P. (2007). Stable isotope evidence for 1500 years of human diet at the City of York, U.K. *American Journal of Physical Anthropology*, 133, 682–697.

Müldner, G., Chenery, C., and Eckardt, H. (2011). The 'headless Romans': multi-isotope investigations of an unusual burial ground from Roman Britain. *Journal of Archaeological Science*, 38, 280–290.

Newman, R. (2006). The early medieval period resource assessment. In *The Archaeology of North-West England: An Archaeological Research Framework for the North-West Region*, Brennan, M., Ed. London, Association of Local Government Archaeological Officers, English Heritage, and Council for British Archaeology, pp. 91–114.

Ottaway, P. (2014). The Ironwork. In *A Post-Roman Sequence at Carlisle Cathedral*, McCarthy, M., Ed. *Archaeological Journal*, 171 (in press).

Paterson, C. and D. Tweddle (2014). Gold, silver, copper alloy. In *A Post-Roman Sequence at Carlisle Cathedral*, McCarthy, M., Ed. *Archaeological Journal*, 171 (in press).

Paterson, C., Parsons, A.J., Newman, R., Johnson, N., and Howard Davis, C. (2014). *Shadows in the Sand: The Excavation of a Viking-Age Cemetery at Cumwhitton, Cumbria*. Lancaster, Oxford Archaeology North.

Pollard, A.M., Pellegrini, M. and Lee-Thorp, J.A. (2011). Technical note: observations on the conversion of dental enamel $\delta^{18}O_p$ values to $\delta^{18}O_w$ to determine human mobility. *American Journal of Physical Anthropology*, 145, 499–504.

Pollard, A.M., Ditchfield, P., Piva, E. et al. (2012). Sprouting like cockle amongst the wheat: the St. Brice's Day massacre and the isotopic analysis of human bones from St. John's College, Oxford. *Oxford Journal of Archaeology*, 31, 83–102.

Price, T.D. and Gestsdottir, H. (2006). The first settlers of Iceland: an isotopic approach to colonisation. *Antiquity*, 80, 130–144.

Price, T.D., Frei, K.M., Dobat, A.S. et al. (2011). Who was in Harold Bluetooth's army? Strontium isotope investigation of the cemetery at the Viking Age fortress at Trelleborg, Denmark. *Antiquity*, 85, 476–489.

Rollason, D. (1998). Symeon's contribution to historical writing in northern England. In *Symeon of Durham, Historian of Durham and the North*, Rollason, D., Ed. Stamford, Shaun Tyas, pp. 1–13.

Sharpe, R. (1998). Symeon as pmphleteer. In *Symeon of Durham, Historian of Durham and the North*, Rollason, D., Ed. Stamford, Shaun Tyas, pp. 214–229.

Smyth, A. (1977). *Scandinavian Kings in the British Isles, 850–880*. Oxford, Oxford University Press.

Whitelock, D. (1979). *English Historical Documents Volume 1, ca. 500–1042*. London, Eyre Methuen.

10

Viking-Age Silver in North-West England: Hoards and Single Finds

Jane Kershaw

CONTENTS

ABSTRACT   Silver hoards of Scandinavian character are arguably the most important archaeological source for Viking activity in the North West. They give a clear impression of the vast pools of silver wealth acquired by the Vikings through plunder and exchange, but why were they concealed in the first place? What do they reveal about the uses of silver in Viking society? And, what can they tell us about Scandinavian activity and settlement in north-west England more broadly? In this chapter, I aim to address such questions by examining the hoards' contents and location, as well as their relationship with single finds from the region.

Introduction

In north-west England, as in many other places across the Viking World, the Viking Age was a silver age. The region is home to a series of silver hoards of Scandinavian character, the most recent of which were discovered only in 2011, at Barrow-in-Furness (Cumbria) and Silverdale (Lancashire). Together, the hoards give a clear impression of wealth generated by Scandinavian activity. But their significance extends far beyond their status as Viking treasure troves. As repositories of coins and objects accumulated by the Vikings through plunder, tribute, and commerce, the hoards offer unparalleled insights into the mobility and cultural contacts of their owners, as well as the economic spheres in which the silver circulated. More broadly, the hoards also illuminate the symbolic value of silver in the Viking World, particularly as a medium for the display of wealth and status.

The aim of this chapter is to provide an overview of current thinking about Viking-Age silver hoards from the North West, as well as to introduce some new research relating to the region's single finds. Accordingly, it addresses several key questions, including: what are the origins of Viking-Age silver from north-west England? What was silver used for? Why were so many hoards deposited in this region? And, what does the combined evidence from hoards and single finds tell us about broader patterns of Scandinavian settlement? But before we can address these questions, we need to have an understanding of the economic activity that preceded the Viking presence. We therefore turn first to consider evidence for wealth and currency before the Vikings.

Background: Wealth and Currency in North-West England before the Vikings

Silver coinage was issued on a large scale in Anglo-Saxon England from the mid eighth century, but evidence for its use in north-west England prior to the arrival of the Vikings is extremely limited. In the ninth century, much of modern-day Lancashire and Cumbria fell under Northumbrian rule and contemporary coins minted at the Northumbrian capital, York, are occasionally found west of the Pennines. However, these consisted not of regular silver pennies of the type minted in other Anglo-Saxon kingdoms, but of small, highly debased silver/copper-alloy coins, known today as *stycas*. The issue of debased coins suggests that precious metal was in short supply in Northumbria, although it is also possible that the coins were specifically designed for use in everyday, low-value transactions (Williams 2008, 48). In the North West, stycas have been found in substantial quantities at the Cumbrian monastic centres of Carlisle and Dacre, and in smaller numbers on coastal sites including Grange-over-Sands (Cumbria); a probable styca hoard, of uncertain size, is also known from Otterspool (Lancs) (Howard-Davis et al. 2009, p. 686; Newman 2006, p. 105; Metcalf 1960, p. 94, pp. 97–98). This distribution points to market activity at settled monastic communities and to seaborne trade along the Irish Sea coast, but is not indicative of a widespread monetary economy. Further south, evidence for coin use in the pre-Viking period in Cheshire, part of Anglo-Saxon Mercia, is even rarer. With the exception of a small group of eighth- and ninth-century coins, including four stycas, from the trading site of Meols on the Wirral peninsula, contemporary coin finds from Cheshire are practically non-existent, meaning that the local population will have been largely unfamiliar with coinage (Griffiths et al. 2007, p. 343).

Coin use thus appears to have been limited in the region, but this is not to say that there was a lack of wealth or indeed currency. Rather than operating a monetary economy, it is likely that the inhabitants of north-west England valued and traded wealth via a different medium, namely commodities (Skre 2011; Gullbekk 2011). It is difficult to gauge the precise nature and extent of commodity exchange from the archaeological record, but surviving documentary evidence from neighbouring regions of the western British Isles indicates a system in which payments could be made in various kinds of goods, on a par with coinage. In the early medieval Irish law codes, for instance, commodities such as cattle, silver and, less certainly, grain are described as means of payments for fines and other forms of social obligations. They could be valued in standardised units relative to each other and also provided units of account (Gerreits 1985). Thus, in the text known as *Crith Gablach*, one *cumal* (a measurement standard, which also means 'female slave') is worth ten cows, and is also a measurement of silver (Charles-Edwards 1993, pp. 478–85). As it appears in the law codes, payment in commodities relates only to social obligations, but it is possible that such a system also characterised commercial trade at local and regional levels (Skre 2011, p. 68). Unfortunately, there is no surviving documentary evidence relating to north-west England, but the practice there of a similar 'commodity money' system seems likely.

The 'Age of Silver'

The nature and longevity of the commodity exchange system in north-west England is difficult to measure. What is clear is that the onset of Scandinavian activity and settlement in the region (beginning, in earnest, in the early tenth century) brought about a tremendous increase in the supply of silver and, with it, a new medium of wealth and currency (Graham-Campbell 1998, pp. 107–16). From this date, a series of silver hoards testify to the operation of a Viking-style silver economy, of a type similar to that in the Scandinavian homelands. The most famous of these was discovered in 1840 on the banks of the River Ribble, close to Cuerdale Hall near Preston, Lancashire (Figure 10.1). The Cuerdale hoard, deposited in c. 905–10, is the largest and most varied silver hoard in the Western Viking World. Containing c. 7,500 coins and over 1,100 extant items of assorted bullion, it weighs an estimated c. 42.6 kg silver

FIGURE 10.1   A selection of artefacts from the Cuerdale hoard, Lancashire (© British Museum).

(Graham-Campbell 2011). To put this in context, the next largest silver hoard from the British Isles after Cuerdale, found at Skaill, Orkney, weighs just over 8kg.

Yet the impressive size of the Cuerdale hoard should not obscure the fact that north-west England is also home to fifteen other precious-metal hoards of broadly contemporary date, making this region the most silver-rich area of Viking England. The remaining hoards vary in size, date and character, and encompass both antiquarian finds and modern discoveries, typically made by metal-detector enthusiasts. While some contain only ornaments, and are thus difficult to date precisely, others can be dated on the basis of coins contained within them (the latest coins in the hoards determining the date after which the hoard was deposited). In north-west England, the series starts with the Cuerdale hoard, deposited in c. 905–10. It ends with a hoard from Halton Moor, Lancashire, deposited in c. 1025, although most hoards fall within the first three decades of the tenth century. This chronological clustering is an interesting trend, and may correlate with a period of pronounced political instability (see below p. 159).

The geographical distribution of the hoards also reveals some interesting trends (Figure 10.2). Many are located along important east-west communication routes connecting the two main centres of Scandinavian power in Ireland and England: Dublin and York (Williams 2009, p. 78). The Cuerdale hoard, for instance, is located close to a north/south crossing of the River Ribble: a position which leads, to the west, to the Ribble estuary and the Irish Sea, and, to the east, to a trans-Pennine Roman road network passing through the Aire Gap into the Vale of York (Graham-Campbell 2011, p. 155, Figure 9.3). Clusters of hoards mark other important crossing points, for instance, at Chester and, further north, at Penrith. A number of hoards, including two recent discoveries from Barrow-in-Furness and Silverdale, populate the lands surrounding Morecambe Bay, which leads, via the River Kent, to a further route to York.

The silver contained within the hoards comes from a mix of sources. Through their raids and trading activity, the Vikings acquired silver coin and other precious metal items from the Islamic Caliphates, the Carolingian Continent, Anglo-Saxon England, and the Danelaw (the area of Scandinavian settlement in northern and eastern England, broadly corresponding with East Anglia, Lincolnshire and Yorkshire). In most parts of Scandinavia, the Vikings did not use coins as coins, in the way that we would today (i.e., by trusting them at face value), but valued them simply for their bullion content. Only the weight and silver purity of the coin was important. Consequently, they preserved some of the silver in its original

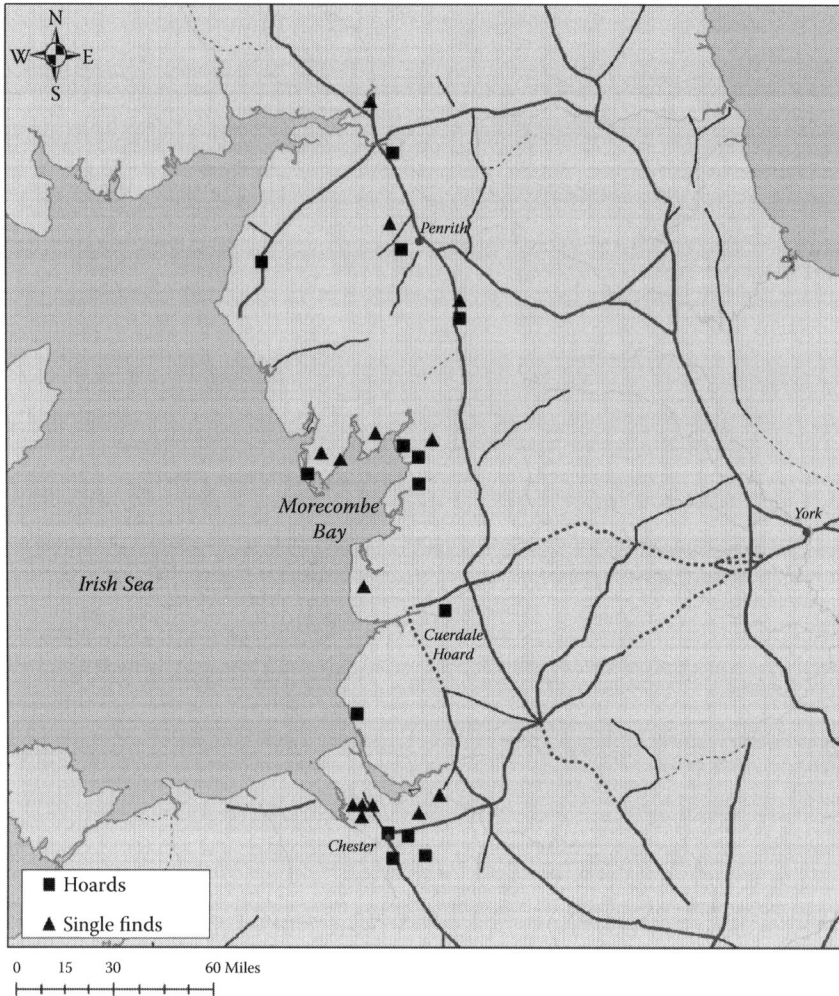

FIGURE 10.2 Hoards and single finds in north-west England shown against Roman roads and major rivers, excluding an unlocalised hoard from 'Lancashire' (© author).

form, but melted down other acquisitions into forms more suitable for the storage, transport and, in some cases, display, of silver, such as ingots and various types of ring. Trace element analysis can sometimes be used to reveal the source of the silver stored in such forms. Metallurgical examination of objects contained in hoards from southern Scandinavia has shown that they were principally made from Arabic silver coins, known as dirhams, but in north-west England, the picture is more varied (Hårdh 1976, pp. 110–27; Arrhenius et al. 1972–3). Ingots from the Cuerdale (Lancs), Scotby (Cumbria) and Castle Esplanade, Chester, hoards do not correlate with any single source of silver. They derive instead from a mix of different sources, and were probably made from silver derived from multiple different coinages and bullion objects (Kruse and Tate 1992).

In hoards from the North West, silver thus takes one of two forms: coins, minted in various countries, and other silver objects, such as jewellery and ingots. Both can be studied to reveal the external cultural contacts of the Vikings, and the contexts of their wealth accumulation. Turning first to coins, what is most striking is the sheer range of different mints represented in the hoards. Although no coinage was minted in north-west England, the Vikings who settled in the Danelaw did produce their own coinage from the late ninth century, and these Viking Danelaw issues are well represented in the region's hoards. Official coins from York, in the northern Danelaw, were produced on a large scale and are particularly

FIGURE 10.3 Islamic silver dirhams are easily recognisable by their large size and Arabic script. This example was found in 2011 in a hoard from Barrow-in-Furness, Cumbria (© PAS).

prominent in hoards from the North West, appearing in the Cuerdale (Lancs), Dean (Cumbria), Harkirk (formerly Lancashire, now Merseyside) and Chester, St John's, hoards. Indeed, York coins are the largest coin group within the Cuerdale hoard. Many were freshly struck at the time of their deposit, suggesting that this coin component had been assembled in York shortly before making its way to Cuerdale (Williams 2011, pp. 43, 70–71).

Perhaps unsurprisingly, Anglo-Saxon coins also contribute a major element in the region's hoards. Of the c. 7,500 coins contained in the Cuerdale hoard, around 900 are in the name of Alfred of Wessex (871–899), although some are copies from the Danelaw rather than official issues. Smaller numbers are recorded in the name of other Anglo-Saxon rulers, including Ceolwulf II of Mercia (874- c. 879) and Edward the Elder (899–924) (Williams 2011, p. 42). Arabic dirhams, large silver coins minted in the Islamic Caliphates, are also contained in the region's hoards (Figure 10.3). The Vikings obtained these coins in exchange for furs, slaves, timber and amber in Russia: huge quantities were imported into Scandinavia, and from there, a smaller number reached Britain and Ireland. Around fifty dirhams are known from the Cuerdale hoard, from mints as diverse as Baghdad in the Abbasid Caliphate, Al-Banjhir in the Hindu Kush, and Al-Andalus in Arabic Spain (Lowick 1976). Dirhams also appear in hoards from Warton (Lancs), Dean and Flusco Pike (Cumbria). Carolingian coins, most likely acquired through Viking raiding and trading in France, are also present in smaller quantities, in hoards from Cuerdale, Harkirk and Dean (Williams 2009, p. 74, Figure 8.1).

Coins thus make up a significant component of silver contained in hoards from the North West. The other main category of hoarded silver is assorted bullion objects, or what is sometimes referred to as non-numismatic silver. These objects include items such as ingots, rings and brooches. Like the coins, they reflect the contacts of the Vikings, most notably with Scandinavia and Ireland. Simple cast bar ingots with oval or D-shaped sections are found in a number of local hoards, including those from Cuerdale and Silverdale (Lancs), Flusco Pike (2) (Cumbria) and Huxley and Eccleston (Cheshire) (see Figures 10.1 and 10.5). They were made by casting molten silver into damp sand or open stone or clay moulds, two examples of which have been found in Chester, at Lower Bridge Street and Cuppin Street (Mason 2007, Figure 30; Bean 2000, p. 17). Cast bar ingots occur throughout Scandinavian-settled regions and are not culturally diagnostic. Nonetheless, it is often assumed that those contained in the Cuerdale hoard originated in Ireland, since this is the likely source of much of the hoard's other bullion content (Kruse 1992, 81; but see Williams 2011, pp. 70–71).

Other types of silver object have identifiable origins. Dublin is thought to have been the main centre of production for a distinctive type of penannular arm-ring (the so-called 'Hiberno-Scandinavian broadband' arm-ring, Hiberno-Scandinavian referring to Scandinavians established in Ireland) (Sheehan 2011). Examples of these arm-rings occur in hoards from Cuerdale and Silverdale (Lancs) and Huxley (Cheshire) (Figure 10.4). Silver penannular (open ring) brooches with long pins, of the type found in the Cuerdale and Flusco Pike (1) hoards, were also produced by Irish-Scandinavian communities, and reflect pre-Viking Irish metalworking traditions. Such objects reached north-west England by way of the Irish Sea. Other types of silver ornament, including arm- and neck-rings made from twisted or plaited rods, are more likely to have been produced within Scandinavia. A silver twisted-rod neck-ring deposited in a hoard from Halton Moor (Lancs) has western Norwegian parallels (Graham-Campbell 2011, p. 89). Spiral rings, so called because of their spiral-striated appearance, were imported from Russia and southern Scandinavia, and are distinctive elements in the Cuerdale and Silverdale hoards. The routes by which such items travelled from Scandinavia to north-west England probably varied: they could have

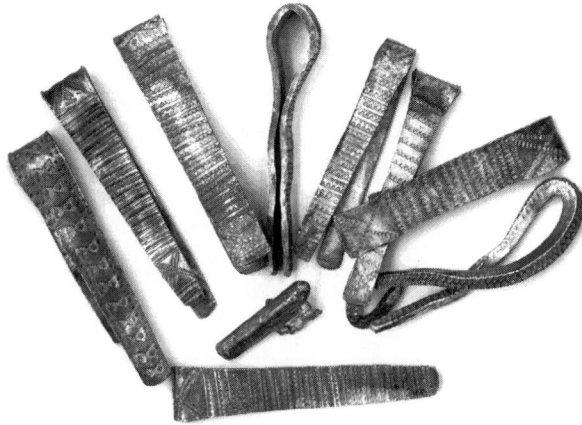

FIGURE 10.4   Folded Hiberno-Scandinavian arm-rings and ingot from the Huxley hoard, Cheshire (© Liverpool Museum).

entered via the Irish Sea, or have travelled westwards through the Danelaw, possibly via Lincoln or York. Combined, then, the coins and bullion objects portray the great diversity of contacts and breadth of silver sources of the Scandinavians in the North West, whilst emphasizing particularly strong connections to the Ireland (in terms of bullion objects) and the northern Danelaw (in terms of coins).

The Uses and Meanings of Silver: The 'Bullion' and 'Display' Economies

The Viking silver economy was thus based on a mixture of distinct artefact types and coin. But why was silver important, and how was it used? Within Scandinavia, silver fulfilled two important roles, and the same is true in north-west England: it had a currency role, and could pass by weight as a means of exchange within a bullion or metal-weight economy, but it also functioned as a status symbol, and could be worn and displayed in shows of wealth, or given as a gift to reward followers and/or to create allegiances and friendships. In this sense, it is possible to speak of two different types of silver economy: one 'bullion' and the other 'display', reflecting the monetary and social roles of silver respectively (Williams 2009, pp. 74–75).* Gold could likewise be used as both a means of payment and of display, although it appears to have been predominantly used in the latter sense.† Gold objects are occasionally found in Viking-Age hoards from Britain and Ireland, but their occurrence in north-west England is rare. The current corpus includes just six gold discs contained in the Halton Moor (Lancs) hoard, and a handful of gold finger-rings recovered as single finds from Sedburgh and Workington (Cumbria), and Aldersey and Chester (Cheshire) (Ager 2011a, pp. 127–28; Graham-Campbell 2011, pp. 107, 160, Handlist 2).

The diverse role of silver in north-west England is reflected in its varying treatment both within and between the region's hoards. Within the Scandinavian bullion economy, silver (and gold) passed by weight and fineness regardless of its form. The deliberate cutting of precious metal provided convenient units of payment, which could then be weighed to the required sums using hand-held balances and weights (some examples of which have been found in the North West, as we consider below). One of the hallmarks of bullion use is therefore deliberately fragmented or 'hack-' silver. In hoards from our region, it is typical to find ingots cut at one or both ends. Some examples bear transverse hammer marks and deep grooves, both of which were designed to prepare the surface for cutting. Brooches and rings were also deliberately fragmented to generate hack-silver, as were Arabic dirhams (Figure 10.5). Other types of coin, for instance, from Anglo-Saxon England and France, are not usually cut, but this probably

---

* In some Scandinavian-settled territories, such as the Danelaw, it is also possible to speak of a third type of silver economy, namely a monetary economy, based on officially-issued coin. However the Vikings did not mint their own coins in north-west England in the ninth or tenth centuries (Williams 2009, 75).

† The will of Wulfric Spot, who held lands in north-west England in around 1000, describes payments to the king in gold, as well horses and weapons (Griffiths 2010, p. 54).

FIGURE 10.5 The Silverdale hoard, Lancashire, contains many deliberately fragmented ingots and ornaments (© PAS).

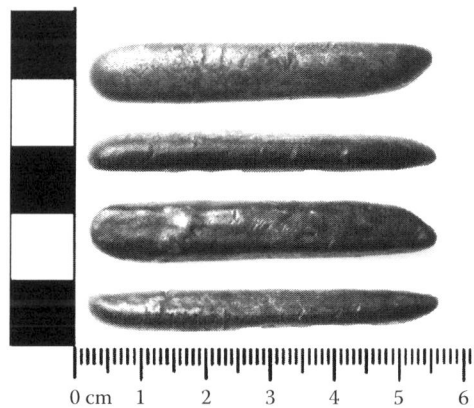

FIGURE 10.6 This ingot from West Yorkshire (just outside our region) displays small knife cuts or 'nicks', indicating it has been tested for its silver content (© PAS).

reflects the fact that they were smaller and lighter than dirhams, and thus already in a form suitable for use in small-scale transactions (Williams 2009, p. 78).

Another physical indication of the use of silver as currency is provided by the appearance of test marks on both coins and non-numismatic objects. Testing the silver was designed to expose plated forgeries and/or to test the fineness of silver by means of a resistance test; an object or coin was thus most likely to acquire test marks when it changed hands during a commercial transaction. Small knife cuts known as 'nicks' were made to the surface of ingots and ornaments, and are visible today as small, crescent-shaped marks (Figure 10.6). Coins were also tested, by means of sharp, angled 'pecks', made by the point of a knife. These survive as small sprues of metal. Interestingly, 'pecking' seems to have been introduced by Vikings in England during a period in which both debased and good quality Anglo-Saxon coins were in circulation, suggesting that the phenomenon arose out of the need to distinguish bad coins from good (Archibald 2011, p. 64). In both forms of testing, if the silver was felt to be too soft, there was a chance it had been adulterated with lead; too hard, and additional copper might be present. Since it probably required quite a lot of experience and skill to judge the 'right' feel of the silver, testing may have been a job for specialists (*ibid*, 56).

A hoard discovered in 2011 in Silverdale (Lancs) is a good example of a predominantly 'hack-silver' hoard, in which the silver seems to have been primarily treated as currency. The hoard was found in a lead container, deposited face down. It contained a mix of silver ingots and ornaments, as well as Anglo-Saxon, Danelaw, Carolingian, and Arabic coins, on the basis of which it has been preliminary dated to

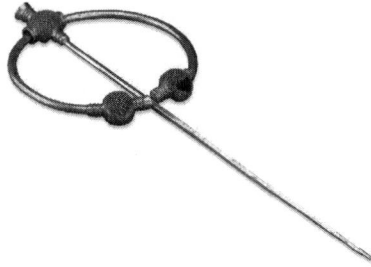

FIGURE 10.7   A penannular 'thistle-brooch' with long pin, Flusco Pike, Cumbria (© British Museum).

c. 900–910. The non-numismatic silver is predominantly in the form of hack-silver, derived from both ornaments and ingots (Figure 10.5). Of 129 ingots in the hoard, for instance, only thirteen are complete, with the remainder cut at one or both ends. Around 75% of the ornaments, chiefly arm-rings, also occur in hack-silver form. Much of the silver in the hoard has also been tested, and many of the coins are bent and/or pecked. Notably, coins minted by Viking rulers from the Danelaw were pecked alongside other types of 'foreign' coin, suggesting that they were not trusted any more than Frankish, Anglo-Saxon or Islamic issues.

Other items in Viking-Age hoards from our region are preserved as complete, intact objects. They highlight an altogether different function of silver in the North West, namely, that of display. Ornaments such as the massive 'thistle-brooches' found at Flusco Pike (Cumbria), so-called because of the thistle-like appearance of their terminals, were prominent status objects: they served as cloak fasteners and were worn on the shoulder with the pin facing upwards (Figure 10.7). Their weight and design made them impractical for everyday dress (the largest 'thistle-brooch' weighs over 700g, with a pin measuring 50cm in length) so it is likely that they were worn only on special occasions, for instance, at public or ritual functions. Such items would have enhanced the status of their owner, but they could also be given as gifts to reward followers and create allegiances. In the Icelandic sagas, kings sometimes give gold rings to members of their retinue as rewards for military service, or to honour court poets (Ager 2011b, pp. 127–28). Perhaps the brooches from Flusco Pike were bestowed for similar purposes. Related brooches are known from the Orton Scar (Cumbria) and Cuerdale (Lancs) hoards, but display objects could also encapsulate items such as complete vessels and weaponry. The Halton Moor hoard, for instance, contained an antique silver-gilt Carolingian cup, which may have functioned as elaborate tableware (Graham-Campbell 2011, cat. no. 4; Wamers 2011, p. 134).

It is important to note that the use of precious metals in these contexts was fluid and flexible. While complete ornaments were primarily intended for display, they also acted as a store of bullion and could be cut up as necessary to generate payment. Indeed, complete objects sometimes carry conspicuous 'nicks', which can be distinguished from scratches and dinks that an object might naturally acquire after a prolonged period of use or in the ground. These indicate that they have been tested for their silver content, as is the case with most of the (more or less) intact silver penannular brooches contained in the Flusco Pike (1) (Cumbria) hoard (Graham-Campbell 2011, cat. nos. 2:1–5). Similarly, fragmented silver could be pooled together and melted down to create new items of jewellery, suitable for display. Although there is no direct physical evidence for this from north-west England, such a process is vividly described in an episode recounted in a thirteenth-century Icelandic saga collection known as *Heimskringla*. In this, the tenth-century poet, Eyvind, receives a reward of silver coins, which is purified and worked into a shoulder-pin weighing 25 lbs (equivalent to over 10 kg of silver, a weight that would have been much too heavy to wear). Rather than wearing the pin, however, Eyvind breaks up the silver, and uses the bullion to purchase a farm (Graham-Campbell 2007, p. 216). Notably, several hoards from our region, including those from Cuerdale and Silverdale, contain both complete ornaments and hack-silver, highlighting co-existence and overlap between the 'display' and 'bullion' economies.

The merging of the display and bullion functions of silver is also attested by ornaments and ingots manufactured to standardised weights. This is the case with several types of arm-ring, for instance, including Hiberno-Scandinavian broad-band arm-rings, complete examples of which reflect a weight unit (c. 26.15g) which seems to have been commonplace in Viking-Age Dublin (Sheehan 2011, p. 99; Wallace 2013, p. 304). Such weight adjustment would not be necessary if the objects were intended only to be worn, but would make sense if they were also to be traded or stored as countable wealth. Complete ingots appear to reflect a similar, but slightly lighter weight unit of c. 25g: a standard more apparent in Scandinavian material (Kruse 1988). A type of plain penannular ring known as 'ring-money', which appears in the Cuerdale hoard, seems to have been made to a target weight of c. 24g (Warner 1976, but see Kruse 1988, p. 288). In a region exposed to influences from both Scandinavia and the Irish Sea, it is perhaps unsurprising that weight units varied across object types. The existence of multiple weight standards will have required traders to be sufficiently flexible to weigh to different units. To ensure confidence in transactions, and to guard against fraud, it is likely that both trading partners measured the silver to be exchanged, using their own hand-held balances and scale weights.

Economic Development

Since silver occurred in a number of different forms, reflecting different types of economy, the composition of hoards can be studied to reveal changes in the use of silver over time. In general within the Scandinavian silver economy, it is possible to identify a gradual transition from the 'display' to 'bullion', and eventually to 'coin' economies, as the monetary functions of silver gradually replaced its symbolic role in social contexts and as the exchange of small sums of hack-silver paved the way for coinage. Certainly, datable hoards within Scandinavia show a general trajectory, starting with intact ornaments in the late eighth and early ninth centuries, followed by a period of silver fragmentation in the tenth century, to hoards dominated by coinage in the eleventh century (Hårdh 1996, pp. 104–107).

However, silver-handling traditions varied across different regions, and north-west England was sandwiched in between two areas with distinct traditions of silver use, namely Ireland and the Danelaw. In the Danelaw, coins were minted by Scandinavian rulers from the late ninth century, and bullion and coin co-existed until the use of bullion ceased in c. 930. In Ireland, by contrast, bullion continued in use into the second half of the tenth century, and coin production only began in the late 990s (Blackburn 2011, 124–36). To judge from the nature of surviving hoards, the 'display' economy also appears to have played a more important role in Ireland relative to the Danelaw, although there is evidence for intact jewellery items even in the latest hoard currently known from Yorkshire (deposited in the Vale of York in c. 928) (Sheehan 2007; Ager and Williams 2011). It would be too simplistic to label Ireland a bullion economy and the Danelaw a coin or dual bullion/coin economy, given evidence for the diverse use of silver in both areas. Nonetheless, it does appear that silver was principally used in different ways in the two regions.

It is thus unsurprising that the composition of hoards from north-west England signals multiple and overlapping silver economies throughout the tenth century, reflecting a combination of influences from east and west. Several early hoards from the region contain significant numbers of coins from York and other Danelaw mints, indicating an exposure to the monetary economy of this region. In addition to containing coin, the Cuerdale and Silverdale hoards also contain large quantities of hack-silver, demonstrating that by c. 900, the concept of the use of silver as a means of payment was well established. Despite this, the silver economy of the North West remained mixed. Hack-silver hoards from the 920s are known from Warton (Lancs) and Flusco Pike (2) (Cumbria). Yet intact ornaments continue to appear in hoards throughout the tenth century, and indeed beyond. Ornament hoards such as that from Flusco Pike (1) (Cumbria), dated on typological grounds to the 920s/30s, bear witness to the survival of the 'display' economy (Graham-Campbell 2001, 224). Even the latest hoard of Scandinavian character on record from north-west England, deposited in c. 1025 in Halton Moor (Lancs), retains a display component in the form of a complete silver cup and neck-ring (Graham-Campbell 2011, cat. no. 4.1 and 4.2).

If the transition from the 'display' to 'bullion' economies is far from clear-cut, the status of coins in the North West is even more complex. Although a conventional coin economy was never established in the region, large numbers of imported coins did reach north-west England in Viking hands. Coins from various mints appear alongside non-numismatic silver in many of the region's hoards, often in a form that indicates that they have been tested for their silver content. Since coins from different mints had different weights and variable silver contents, it is likely that they passed primarily as bullion, to be weighed out alongside other forms of silver in metal-weight transactions.

However, from the 910s, several coin-only hoards are also recorded. While two of these, from Dean (Cumbria) and Chester, St John's, contained a mix of coins from multiple mints, a hoard of the same date from 'Lancashire' is formed entirely of York issues (Graham-Campbell 2001, pp. 218–19). It is possible that the owner of this hoard accepted the coins as coinage, passing them by tale (that is, by counting them out) rather than by weight. Although removed from the area of jurisdiction in which they were minted, the York coins may have achieved a nominal status as coins, in effect providing a substitute for a local currency.* Certainly, by the later tenth century, there is evidence that coins in hoards from north-west England were treated differently to silver in other forms. This is perhaps most clearly seen in the recently discovered hoard from Barrow-in-Furness, deposited in c. 955–57. The 79 coins found along with thirteen ingot fragments were preserved either as complete coins or as deliberately cut halves, while only around a quarter had been tested for their silver content. This treatment suggests that the coins were, to an extent, taken at face value, the presence of coin halves speaking of the desire to create smaller units of currency that could still be accommodated alongside complete coins in what were presumably coin-based or mixed bullion/coin transactions. In this way, the hoard would appear to indicate a shift towards the appreciation of coin as coin.

Reasons for Hoarding

Precious-metal hoards represent accumulations of wealth that were deposited by their owners and never recovered. But why did silver enter the ground in the first place, and why did north-west England in particular see hoarding on such a large scale in the early tenth century? Archaeologists have interpreted hoards in a number of different ways. A common explanation is that they were buried for safe keeping during periods of political turmoil or absence. To take an example from a much later period, Samuel Pepys, the seventeenth-century diarist, gives a vivid account of his concealment of over 1,300 pounds of gold upon hearing about a threatened Dutch invasion. He instructed his father and wife to bury the gold on his family estate in Huntingdonshire, but when they came to recover the hoard a year later, they couldn't remember exactly where it had been buried. It was only 'by poking with a spit' that Pepys eventually located the hoard, and even then he found that the coins had been scattered and some lost for good (Latham and Matthews 1985, pp. 788–89, 838–39).†

In other cases, hoards may have been concealed without a specific threat, simply for the purpose of storing value and stockpiling wealth for future use. Here, hoards could be viewed as proxy savings banks (without the benefit of interest). Hoards containing objects spanning a long date range are candidates for such deposits, since this suggests that they were added to over an extended period of time (Sindbæk 2011, p. 57). In both contexts, the hoards known to us today represent 'failed' hoards that were never reunited with their owner. If we assume that most hoards were successfully recovered, those that 'survive' must represent just a small fraction of the number of hoards that were originally concealed.

Other hoards may have been deposited without the intention of recovery, for ritual or ceremonial reasons. Viking-Age gold hoards and single gold finds from Scandinavia, Britain and Ireland, are frequently associated with watery places such as bogs, rivers and lakes, from which they would have been difficult (although perhaps not impossible) to retrieve (Hårdh 1996, p. 134; Graham-Campbell and Sheehan 2009). An association with silver hoards and watery environments has not been established, which may suggest a broad distinction in the treatment of hoards containing (silver) currency, and those containing (gold) display items. An altogether different explanation for hoarding is offered by the thirteenth-century Icelandic

---

* It may seem odd that coin could serve as currency outside of the area in which it was minted, since there would be no legal framework guaranteeing its value. However, there are numerous examples of foreign coin serving an important role in local economies, as in Colonial America, when settlers adopted various foreign currencies, including the Spanish 8-reales, or in seventeenth-century Poland, when local coin supply was dominated by Swedish and Saxon coin. Indeed, in tenth-century Ireland, Anglo-Saxon coins were a trusted currency prior to the establishment of a mint in c. 997 (Bornholdt-Collins 2010, pp. 22–23; Eagleton and Williams 2011, p. 167; Sheehan 2007, p. 159).

† The entries are for the 13th June and 10–11th October 1667.

historian, Snorri Sturluson. He states that pre-Christian Scandinavians believed that they would have access to whatever they had buried in hoards in the afterlife, a principle sometimes referred to as 'Odin's Law' (Ager 2011b, p. 133). This belief is also indirectly referenced in the Icelandic sagas. In *Egil's Saga*, for instance, Egil's father deposits a chest of silver and a bronze cauldron in a marsh immediately prior to his death; before he dies, Egil himself hides two chests of silver, again in boggy land (Graham-Campbell and Sheehan 2009, p. 89). It is unclear, however, whether such accounts reflect genuine tenth-century beliefs or simply literary devices to account for the activities of earlier, pre-Christian generations.

Returning to the Viking-Age North West, the contents and contexts of hoards can provide some clues as to the possible motivations for their concealment. In our area, most hoards contain a currency element, and thus are usually accounted for within an economic, rather than ritual, framework. Certainly, the concentration of hoarding in the first three decades of the tenth century does correspond with a period of pronounced political upheaval between Anglo-Saxon, Irish and Scandinavian forces, as Scandinavian exiles from Ireland sought new lands in Cheshire and coastal Lancashire, and Anglo-Saxon rulers from Mercia and Wessex looked to extend their authority into the power vacuum left by the earlier collapse of Northumbria (Griffiths 2010, pp. 41–45).

The landscape contexts of the hoards from our region have not yet been studied in any detail, although they could usefully be explored in the case of the most recent discoveries. However, no hoards are known to derive from watery environments. Here, it is also worth observing that a number of hoards, including those from Cuerdale and Silverdale (Lancs), Huxley (Cheshire) and Dean (Cumbria) were buried in lead containers, while a hoard from Chester, Castle Esplanade, was contained within a pot (Graham-Campbell 2011, pp. 130–31). The use of such containers, which were presumably intended to protect the hoards, may indicate that they were intended for recovery (Ager 2011b, p. 133).

The combined evidence thus suggests that many of the region's hoards are likely to have been deposited for safekeeping in a period of general upheaval, although some hoards may well have been concealed for different reasons. Individual hoards have plausibly been linked with documented 'crisis' events. On the basis of its date, location and strong Scandinavian-Irish characteristics, the Cuerdale hoard has been interpreted as wealth accumulated by the Irish-Norse elite following their expulsion from Dublin and Waterford in 902 (Graham-Campbell 1992, p. 114). The contemporary hoards from Silverdale (Lancs) and Huxley (Cheshire) may belong to a similar context. Two hoards concealed in the 920s/30s in Flusco Pike, near Penrith, may have been connected with the meeting at nearby Eamont Bridge between the West Saxon King, Æthelstan, and the Kings of the Scots and of Cumbria in 927. During this gathering, the northern rulers submitted to Æthelstan, who secured a new northern frontier: events that may have encouraged the safeguarding of wealth (Williams 2009, 80; Graham-Campbell 2011, p. 156). There are, however, difficulties in correlating hoards with specific historical events; it is equally possible that the hoards relate to events which are not known to us today.

Hoards vs Single Finds

Whatever the motivation for hoarding, hoards are distinct from a second category of evidence for the silver economy in north-west England, namely single finds. Whereas hoards represent deliberate deposits of accumulated wealth, single finds are individual items, for instance, ingots or coins, that were most likely dropped or mislaid by their owners. They are typically retrieved during metal-detector surveys and reported to the Portable Antiquities Scheme (PAS), which hosts an extensive on-line database of such discoveries (www.finds.org.uk). As a source of information for Scandinavian silver economies, single finds are potentially hugely valuable. Unlike hoards, they are not confined to precious metal, but also include tools used in commercial transactions, such as weights and balances. Moreover, whereas hoards are deliberately concealed deposits, single finds are assumed to represent accidental losses, and will therefore reflect a more or less random sample of silver and related artefacts in circulation.

Given the large number of hoards recovered from the region, we might expect single finds to be similarly plentiful. In fact, this is not the case. The corpus of single finds of silver bullion is small, comprising just a cut fragment of an Arabic dirham, from Skelton (Cumbria); a Frankish silver coin of Charles the Bald (840–77), found in Puddington (Cheshire); and an ingot with 'nicks' from Aston (Cheshire) (PAS

FIGURE 10.8   A decorated lead weight, Newton, Lancashire (© PAS).

'Find-ID' LVPL522; Cowell and Philpott 1993; *Treasure Annual Report* 2003, no. 83). A further nicked ingot, together with a piece of hack-silver, is recorded from the village of Neston near the Dee estuary (Griffiths 2010, p. 115, Figure 68; Bean 2000). A silver Thor's hammer, which may have been worn as a pendant or used in a bullion context, is also recorded, from Longtown (Cumbria) (PAS 'Find-ID' LANCUM-ED9222) (a further Thor's hammer find, said to come from Cumbria, was recently dismissed as a fake). In addition to these silver items, several gold finger-rings of Scandinavian type are on record (see above). These rings are complete, and thus in a form that would appear to relate primarily to the 'display' economy.

Somewhat more common as single finds, totalling seven items, are lead weights decorated with inset metalwork (see, for instance, PAS 'Find-ID' LANCUM-45FF34; LVPL1049) (Figure 10.8). This is a fairly common type of Viking scale weight, thought to be associated with the weighing of bullion: examples have been found with hand-held balances in Viking-Age graves, and at market and productive sites in presumed commercial contexts (Graham-Campbell 1980, no. 307; Redknap 2009, p. 38). Just outside of our region, two decorated lead weights have been found at the fortified coastal trading-site of Llanbedrgoch, Anglesey, and have been linked to merchant activity (Redknap 2000, p. 61, Figure 82). The added metalwork is usually British or Irish in origin, and thus the weights have been identified as an Insular Viking phenomenon, created mainly during the later ninth century, but remaining in production and use into the tenth. The exact function of the added metalwork is unclear, but one possibility is that it served to personalise weights, allowing the owner to easily recognise their set in a trading environment where multiple sets of weights were in use.

The small number of single finds makes them difficult to evaluate as a group. Nevertheless, their findspots have the potential to reveal locations in which bullion-users were concentrated. Figure 10.2 shows that the single finds cluster in two areas: around the Dee and Mersey estuaries flanking the Wirral, and on the coasts surrounding Morecambe Bay in north Lancashire and southern Cumbria, including on the Furness peninsula. Notably, both areas also see concentrations of hoards, although there is no direct overlap between the two find categories. It is unclear if the single finds represent stray losses made by people travelling across the landscape, or if they relate to permanent settlements or market sites that may have had a more seasonal character. Certainly, the Wirral peninsula is considered by many to have been a focus of Scandinavian settlement and the single finds may relate to concentrations of bullion users at its southern edge (Philpott, Chapter 7). Morecambe Bay was also an important, strategic harbour, and a number of burials and stone sculptures from its coastal edges hint at significant local Scandinavian landholding (Griffiths 2010, pp. 55–56, Figure 19). Metal-weight exchange may thus relate directly to Scandinavian settlement in these two areas.

Overall, the number of single finds from the North West remains small. Notably, the area of the Danelaw has yielded a far greater number of single finds, including over 60 ingots, 50 dirhams and eight neck- or arm-rings/ring fragments. Also found in the region are numerous lead and copper-alloy weights,

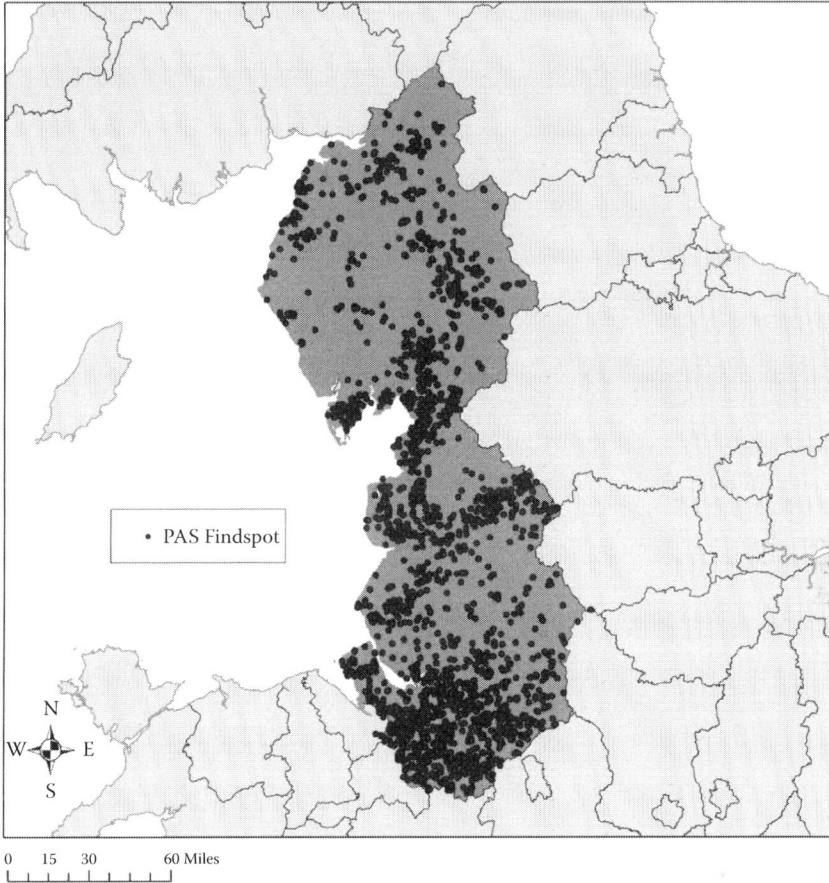

FIGURE 10.9 Roman, Medieval and Post-Medieval finds recorded by the PAS in north-west England as of July 2013 (© author).

including types not recorded at all in the North West. What factors may explain this difference? Metal-detecting is certainly more common in eastern England than in the west, where much land is unavailable to detectorists because of large urban centres and extensive pastoral farming. However, north-west England, particularly Cheshire and the area of southern Cumbria and northern Lancashire, has produced a significant number of portable antiquities from the Roman, Medieval and post-Medieval periods (Figure 10.9). Metal-detector coverage is not, then, the cause of the low number of bullion-related single finds from the region. Indeed, if we look at other types of early medieval finds such as dress accessories, we find that they are also rare, a pattern which suggests that the frequency of bullion-related finds in particular is representative of broader patterns in contemporary metalwork consumption and loss (Richards et al. 2009, 3.3.1.2). North-west England is a region of highland, and was historically sparsely settled (*ibid*, 3.3.1.1). Even with an influx of Scandinavian settlement in the early tenth century, it is likely that overall population levels remained low.

Why, then, has the north west yielded so many hoards? It is possible that hoards have a higher chance of being detected and recorded than single finds, both in antiquity, and in modern-day metal-detecting. However, in other parts of England we get the opposite pattern: East Anglia, for instance, has yielded a vast number of single finds, but no certain hoards of Scandinavian character. An arguably more likely explanation for the pattern in north-west England is that the hoards were simply passing through the region at the time they were concealed. Rather than reflecting local silver use, the hoards probably represent wealth in transit. Of course, there may well be exceptions, and, as noted above, individual hoards have been linked to centres of Scandinavian land holding in the region. Nevertheless, the hoards are often located on prominent east-west communication routes linking the northern Danelaw with the Irish Sea region: a pattern that highlights the importance of the North West as a conduit for the movement of wealth, and presumably also people and goods, between the two regions. Despite the richness of the region's hoards, the overall conclusion must be that north-west England was primarily an area through which silver travelled, rather than one in which it was actively used.

Conclusions

The hoards of north-west England reveal the silver riches of the Vikings active in the region during the tenth century, but they are more than showcases for Viking plunder and tribute. Not only do they reveal the external contacts and wealth sources of the Scandinavians, they also elucidate the ways in which the Vikings conceptualised and valued silver, in both the economic and symbolic spheres. As I hope to have demonstrated in this chapter, the origins of the hoarded silver are diverse, reflecting the background of Vikings in the Scandinavian homelands, as well as their activities in Russia, on the Continent, in England (including in the Danelaw) and in the Irish Sea region. Overall, however, the hoards show strongest connections with Ireland and the northern Danelaw, highlighting the prevailing influence of these two areas over north-west England.

The uses of silver in the region were similarly diverse, the hoards demonstrating the role of silver both as a means of payment within a metal-weight system, and as a means of display in social settings. There is little evidence for changes in the use of silver over time. Although we might expect the 'display' economy to give way to the 'bullion' economy, it seems instead that the uses of silver remained mixed for much of the tenth century, a pattern which serves as a reminder that silver economies developed in different ways in different parts of the Viking World. The reasons why so many hoards were deposited in this area of England are opaque, but a case was made for most hoards being deposited for economic reasons, at a time of general political instability. What is made clear from a comparison with the region's single finds is that the hoards are anomalous in their local context. Rather than representing silver that was in active use locally, the hoards appear to represent wealth in transit between York/the northern Danelaw and Dublin/ the Irish Sea region. In this sense, the hoards demonstrate the movement of wealth between east and west, and highlight the role of north-west England as an intermediary link between the two areas.

REFERENCES

Abramson, T. (2011). *New perspectives. Studies in Early Medieval Coinage*. Woodbridge: Boydell & Brewer

Ager, B. (2011a). The Halton Moor gold discs. In Graham-Campbell, 127–28.

Ager, B. (2011b). A preliminary note on artefacts from the Vale of York hoard. In Abramson, ed, 121–134.

Ager, B. and G. Williams (2011). The Vale of York hoard: preliminary catalogue. In Abramson, ed, 135–145.

Archibald, M.M. (2011). Testing. In Graham-Campbell, 51–64.

Arrhenius, B., U.S. Linder Welin and L. Tapper (1972–3). Arabiskt silver och nordiska vikingasmykken. *TOR* 15, 151–159.

Bean, S. (2000). Appendix. Silver ingot from Ness, Wirral. In *Wirral and its Viking Heritage*, eds. P. Cavill, S.E. Harding and J. Jesch, 17–18, Nottingham: English Place Name Society.

Blackburn, M A.S. (2011). Currency under the Vikings. Part 5: The Scandinavian achievement and legacy. In *Viking Coinage and Currency in the British Isles*, M. Blackburn, 119–147. London: Spink.

Bornholdt Collins, K. (2010). The Dunmore Cave [2] hoard and the role of coins in the tenth-century Hiberno-Scandinavian economy. In *The Viking Age: Ireland and the West*, eds J. Sheehan, D. Ó Corráin and S. Lewis-Simpson, 19–46, Dublin: Four Courts Press.

Charles-Edwards, T. (1993). *Early Irish and Welsh Kinship*. Oxford: Clarendon Press.

Cowell, R. and R. Philpott (1993). Some finds from Cheshire reported to Liverpool Museum. *Cheshire Past* 3, 10–11.

Eagleton, C. and J.D. Williams (2011). *Money: A History*. London: British Museum Press.

Gerreits, M.E. (1985). Money in early Christian Ireland according to the Irish laws. *Comparative Studies in Society and History* 27, 323–339.

Graham-Campbell, J. (1980). *Viking Artefacts: A Select Catalogue*. London: British Museum Publications.

Graham-Campbell, J. (1992). The Cuerdale hoard: comparisons and context. In Graham-Campbell, ed, 107–115.

Graham-Campbell, J., ed. (1992). *Viking Treasure from the North West: the Cuerdale Hoard in its Context*. Liverpool: National Museums and Galleries on Merseyside.

Graham-Campbell, J. (1998). The early Viking Age in the Irish Sea area. In *Ireland and Scandinavia in the Early Viking Age*, eds H.B. Clark, M. Ní Mhaonaigh and R. Ó Floinn, 104–130, Dublin: Four Courts Press.

Graham-Campbell, J. (2001). The northern hoards: from Cuerdale to Bossall/Flaxton. In *Edward the Elder, 899–924*, eds N.J. Higham and D.H. Hill, 212–229, London: Routledge.

Graham-Campbell, J. (2007). Reflections on 'silver economy in the Viking Age'. In Graham-Campbell and Williams, eds, 215–223.

Graham-Campbell, J. (2011). *The Cuerdale Hoard and Related Viking-Age Silver and Gold from Britain and Ireland in the British Museum*. London, British Museum Publications.

Graham-Campbell, J. and R. Philpott (2009). *The Huxley Viking Hoard: Scandinavian Settlement in the North West*. Liverpool: National Museums Liverpool.

Graham-Campbell, J., S.M. Sindbæk and G. Williams, eds. (2011), *Silver Economies, Monetisation and Society in Scandinavia, AD 800–1100*, Aarhus: Aarhus University Press.

Graham-Campbell, J. and J. Sheehan (2009). Viking-age gold and silver from Irish crannogs and other watery places. *Journal of Irish Archaeology* 18, 77–93.

Graham-Campbell, J. and G. Williams, eds. (2007). *Silver Economy in the Viking Age*. Walnut Creek: Left Coast Press.

Griffiths, D. (2010). *Vikings of the Irish Sea*. Stroud: History Press.

Griffiths, D., R.A. Philpott and G. Egan eds. (2007). *Meols: the Archaeology of the North Wirral coast*. Oxford: School of Archaeology.

Gullbekk, S.H. (2011). Norway: commodity money, silver and coins. In Graham-Campbell, Sindback and Williams et al., eds, 93–111.

Hårdh, B. (1996). *Silver in the Viking Age: a Regional-Economic Study*. Stockholm: Almquist & Wiksell International.

Hårdh, B. (1976). *Wikingerzeitliche Depotfunde aus Südschweden: Probleme und Analysen*. Lund: Liber-Läromedel/Gleerup.

Howard-Davis, C., A. Bates and A. Parsons (2009). *The Carlisle Millennium Project: Excavations in Carlisle, 1998–2001*. Lancaster: Oxford Archaeology North.

Kruse, S. (1988). Ingots and weight units in Viking Age silver hoards. *World Archaeology* 20(2), 285–301.

Kruse, S. (1992). Metallurgical evidence of silver sources in the Irish Sea province. In Graham-Campbell, ed, 73–88.

Kruse, S. and J. Tate (1992). XRF analysis of Viking Age silver ingots. *Proc. Soc. Ant. Scotland* 122, 295–328.

Latham, R. and W. Matthews eds. (2003). *The Diary of Samuel Pepys. A Selection*. London: Penguin.

Lowick, N.M. (1976). The Kufic coins from Cuerdale. *British Numismatic Journal* 46, 19–28.

Mason, D.J.P. (2007). *Chester AD 400–1066: from Roman Fortress to English Town*. Stroud: Tempus.

Metcalf, M.D. (1960). Some finds of Medieval coins from Scotland and the North of England. *British Numismatic Journal* 30, 88–123.

Newman, R. (2006) Early medieval period resource assessment. In *The Archaeology of North West England. An Archaeological Research Framework for the North West Region*, ed M. Brennand, 91–114. Council for British Archaeology North West.

Redknap, M. (2000). *Vikings in Wales. An Archaeological Quest*. Cardiff: National Museum and Galleries of Wales.

Redknap, M. (2009). Silver and commerce in Viking-Age North Wales. In Graham-Campbell and Philpott, eds, 29–41.

Richards, J., J. Naylor and C. Holas-Clark (2009). Anglo-Saxon landscape and economy: using portable antiquities to study Anglo-Saxon and Viking Age England. *Internet Archaeology* 25.

Sheehan, J. (2007). The form and structure of Viking-Age silver hoards: the evidence from Ireland. In Graham-Campbell et al., eds, 149–162.

Sheehan, J. (2011). Hiberno-Scandinavian broad-band arm-rings. In Graham-Campbell, 94–100.

Sindbæk, S.M. (2011). Silver economies and social ties: long-distance interaction, long-term investments—and why the Viking Age happened. In Graham-Campbell et al., eds, 41–65.

Skre, D. (2011). Commodity money, silver and coinage in Viking-Age Scandinavia. In Graham-Campbell et al., eds, 67–91.

Wallace, P. (2013). Weights and weight systems in Viking Age Ireland. In *Early Medieval Art and Archaeology in the Northern World: A Festschrift for James Graham-Campbell,* eds A. Reynolds and L. Webster, 301–316. Leiden: Brill.

Warner, R. (1976). Scottish silver arm-rings: an analysis of weights. *Proceedings of the Society of Antiquaries of Scotland* 107, 136–143.

Wamers, E. (2011). The Halton Moor cup and the Carolingian metalwork in the Cuerdale hoard. In Graham-Campbell, 133–39.

Williams, G. (2008). *Early Anglo-Saxon Coins*. Oxford: Shire Publications.

Williams, G. (2009). Viking hoards of the Northern Danelaw from Cuerdale to York. In Graham-Campbell and Philpott, eds, 73–83.

Williams, G. (2011). The Cuerdale coins. In Graham-Campbell, 39–71.

11

What Can Genetics Tell Us about the Vikings in the Wirral and West Lancashire?

Turi King

CONTENTS

> "Yes, there is perhaps more of Norse blood in your veins than you wot of, reader, whether you be English or Scotch; for those sturdy sea rovers invaded our lands from north, south, east and west many at time days gone by, and held it in possession for centuries at a time, leaving a lasting and beneficial impress on our customs and characters"

> **Erling the Bold**
> *by R.M. Ballantyne (Ballantyne 1869)*

ABSTRACT The Vikings are known to have arrived in the Wirral and West Lancashire over 1000 years ago. Until recently, the evidence of this has come from archaeological, linguistic and textual sources. Recently another source of evidence has presented itself: a genetic signal of their presence may exist in the DNA of those who live in the area today. This paper discusses a pilot project which sought to do just that: investigate if any genetic traces of the Viking migration to the area could be found among the modern population of the Wirral and West Lancashire.

Introduction

There can be few historical-cultural groups that are viewed more romantically by the British public than the Vikings. From the Victorian vision of noble warriors invading the British Isles (even Queen Victoria was claimed to have descended from them) to the Viking re-enactor groups today, they are a much celebrated part of our nation's heritage. Representations are embedded in our modern culture on, among other items, the emblem on the front of Rover cars, beer labels (there is even 'Viking beer') and hotel signs. And a quick flick through the history documentaries on our television screens proves that the fascination with Vikings continues unabated.

The Vikings arrived in Britain over a thousand years ago and, until recently, the evidence had come from the traditional sources of texts, archaeological and linguistic evidence, used to piece together information about their presence and influence in the British Isles. There are very few surviving local contemporary written accounts: the Vikings are mentioned in the Anglo-Saxon Chronicle and Irish Annals and a handful of other sources. These speak of the first Viking contacts as ones of raiding: Wulfstan, the Anglo-Saxon Bishop of London in the late 10th century famously (in his Sermon of the Wolf) saw the Viking raids as God punishing the Anglo-Saxons for their sins (Smail and Gibson 2009). A shift towards trading, migration and later permanent settlement then took place in the mid-ninth century (Richards 2004). The settlers are thought to have integrated with the indigenous population and the influence of this Viking presence, particularly in parts of the country that were under the administration of the Vikings, survives today. Archaeologists find physical evidence of their settlement and daily lives, Scandinavian word elements pervade our language, and numerous place-names containing Scandinavian elements bear witness to the Viking settlers naming the landscape (Barber, et al. 2009). Indeed, many of our place names (such as those ending in *–by*, *-thorpe*, *-toft* and *–thwaite*) and words used in English today are Scandinavian in origin and in areas of the Danelaw, up to ~70% of major place names are Scandinavian in origin.

Vikings in the Wirral and West Lancashire

An area of Britain where the Vikings are thought to have arrived in some numbers is the Wirral and West Lancashire. Here Norse Vikings are thought to have arrived via Dublin (Griffiths 2010) and are often referred to as Hiberno-Norse. Ireland had become an area of intense Viking activity from the late 8th century AD and by 841AD the Vikings had a fortified camp at Dublin (Haywood 1995). Dublin remained as a Viking raiding base until 902AD when a number of sources, including the Irish annals, describe a battle (Griffiths 2010) which led to the expulsion of the Vikings from Dublin. The annals tell of Aethelflaed, Queen of the Saxons, granting the Vikings new lands near Chester, now thought to be the Wirral. A further group is thought to have eventually settled in Lancashire (Fellows-Jensen 1985; Higham 1992).

Certainly the place-name evidence strongly supports Viking settlement in this area: Norwegian place-name elements are common in north-west England (Fellows-Jensen 1985; Fellows-Jensen 1992; Fellows-Jensen 2000; Barber et al. 2009), and place-names with Scandinavian and Hiberno-Norse elements are found at high frequency in the Wirral and West Lancashire (Higham 1992; Cavill et al., 2000). Perhaps even more telling is the presence of the place name Thingwall (from Old Norse *þing-vollr*) meaning assembly field indicating a Norse meeting place or parliament and suggesting a community of sufficient size to warrant one—one of only a handful found in the British Isles. Further evidence of Viking settlement is found in the form of Viking crosses, hog-back tombstones, coins, the archaeological finds on the Wirral at Meols (from *melr* the Viking word for sand-hills), a Viking house at Irby, the Viking hoards of Chester, Huxley, Harkirke, Cuerdale and numerous other small finds in the area.

Thus the archaeological, linguistic, written and place name evidence suggests that the Vikings settled in the Wirral and West Lancashire. As geneticists, we are interested in the genetic contribution of the

Vikings to the population and ask whether we can find genetic traces of the Viking migration in the DNA of people of this area today.

How Can We Use DNA to Find Traces of the Viking Migration to the Wirral and West Lancashire

DNA and Chromosomes

The DNA in each of our cells can be thought of as an instruction manual passed down from parents to their children. This manual provides the body with instructions, in the form of genes, which direct the development, growth and maintenance of our bodies and reproduction, as well as the characteristics (sex, hair colour, eye colour etc.) that each of us has (see, for example, Sulem et al., 2007). As we inherited our instruction manual from our parents, who inherited it from their parents, as so on back, our DNA can be likened to a message from our ancestors.

The majority of the DNA that we each carry in our cells is found in a small circular structure within the cell known as a nucleus, and is known as nuclear DNA (Figure 11.1). Within the nucleus, our DNA is packaged into 46 chromosomes arranged into 23 pairs with one half of each pair coming from our mother and the other half from our father. These 23 pairs themselves can be divided into two different types of chromosomes: chromosome pairs number 1-22 (where each of the pairs of chromosomes are the same size) are known as our autosomes, with these being the same between men and women. Chromosome pair number 23 is our sex chromosomes and it is here that men and women differ. Women have two copies of a large chromosome known as chromosome X. Men also have an X chromosome but instead of a second X chromosome they have a smaller chromosome known as the Y chromosome. The Y chromosome carries on it the gene which determines maleness and therefore the presence or absence of this chromosome determines the sex of an individual: if you have a Y chromosome you are male and if not you are female. The Y chromosome, therefore, can only be inherited down the male line.

Another portion of our DNA, known as mitochondrial DNA(mtDNA), is found in small structures known as mitochondria which are located within the cell, but outside the nucleus. The mtDNA that we inherit is that from the ovum (egg) and therefore mtDNA is only passed down through the female line.

Our chromosomes are essentially a very tightly wound and packaged strand of DNA. The DNA strand is made up of just four building blocks known as bases: guanine (G), adenine (A), thymine (T) and cytosine (C). The DNA strand looks like a ladder with the sides of the ladder being made up of sugar-phosphate strands and the bases pairing together to form the rungs of the ladder. Very importantly there

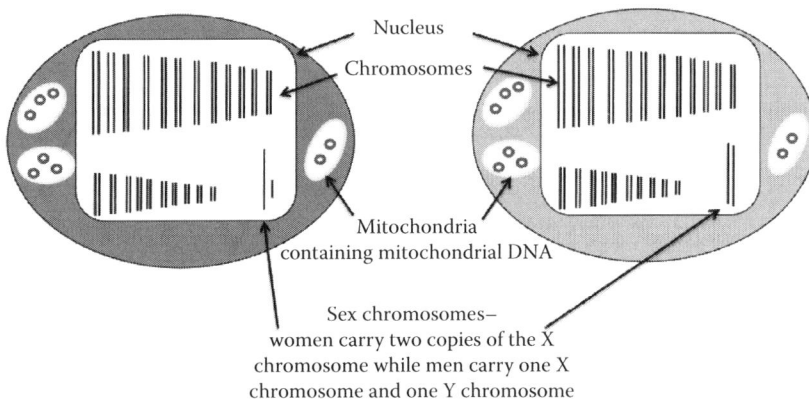

FIGURE 11.1 Our DNA can be divided into two different sorts. These are nuclear DNA, consisting of 23 pairs of chromosomes, and mitochondrial DNA, found in the mitochondria.

is a pairing rule in terms of forming the rungs: G always pairs with C and T always pairs with A. It is the sequence of these base-pairs along the ladder that forms the genetic code and is referred to as the DNA sequence.

The pairing rule facilitates DNA replication during the formation of new cells with each side of the ladder forming the template to build the other side. Furthermore, when writing out a DNA sequence, only one of the strands is necessary. For example, instead of writing

```
AATCTGTAGATCCT
| | | | | | | | | | | | | |
TTAGACATCTAGGA
```

writing AATCTGTAGATCCT is sufficient to know what the full double-stranded sequence will be.

The human genome is made up of ~3 billion of these base pairs, divided up into 23 chromosomes. We have two copies of the genome in our cells (one copy from mum and one from dad) and hence the 23 pairs of chromosomes. Our genes are simply sections of our DNA which code for particular proteins to be made. Our genome is thought to contain ~24,000 genes which account for only ~2% of its content. The function of the remainder of our DNA is the subject of much investigation but some is thought to have a regulatory effect on our genes.

Not all of our DNA is the same from one person to another. All of us have slight differences in our DNA that make us different from one another and they can occur within genes or between them. For example one person may have a very slightly different DNA sequence for one gene than another person which will then code for a different protein. We call these two different forms of the same gene *alleles*. One person may have an allele that causes their eyes to be more blue in colour and another may have the allele that causes their eyes to be more brown. The allele that each person has at a particular DNA sequence is known as their genotype. Sites where the DNA sequence is known to differ between individuals are known as markers. Common types of DNA markers used to distinguish individuals apart and how they arise will be discussed in a later section. Many of the markers that we use in our studies are not found in genes and do not have any effect on any visible characteristics.

Inheritance of the Y Chromosome and Mitochondrial DNA

The Y chromosome and mitochondrial DNA are unusual. They do not undergo a process known as recombination (where each pair of chromosomes line up next to one-another and shuffle information between them) before being passed down to the next generation in sperm and eggs. Where most of our DNA is a patchwork of that of our ancestors and it becomes impossible to tell which piece of DNA came from which distant ancestor, the Y chromosome is passed down virtually unchanged from father to son down through the generations. The changes that do occur are the small mutations, which can be likened to typos, when the Y chromosome (like all chromosomes) is copied to be passed down to the next generation. We geneticists like these little mutations as they introduce variation in the DNA sequence and allow us to tell Y chromosomes apart. Each man's Y chromosome will contain within it a record of all the mutations (the overwhelming majority of which occur in the regions between genes and are harmless) that have occurred along his male-line lineage over time and these will be different from those of the next man's Y chromosome. Mitochondrial DNA is inherited in an equally simple fashion down the female line. Therefore these two pieces are unusual in that they contain relatively simple records of our genetic past—albeit just representing just two of our many ancestors (Figure 11.2).

There are two different commonly used types of marker that we use to tell Y chromosomes apart (Figure 11.3). The first of these are known as binary markers, so called because they come in either one form (allele) or another. The most common type of binary marker that we look at is a Single Nucleotide Polymorphism (SNP—polymorphism meaning many forms). The hundreds of these SNPs now known to exist on the Y chromosome are essentially where there is a change in the base found at a particular location on the Y chromosome.

These types of marker have a very low mutation rate and once such a mutation has occurred in a man's lineage then this will be passed down to his male-line offspring. These types of mutations happen so

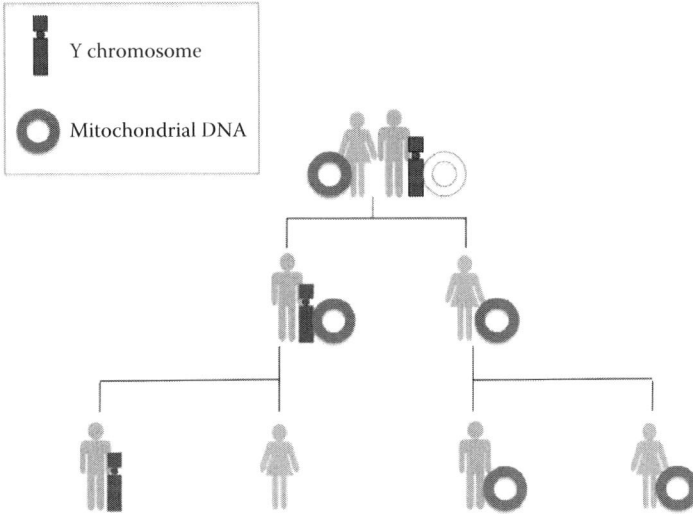

FIGURE 11.2   The Y chromosome and mitochondrial DNA are two sections of our DNA which have a simple pattern of inheritance. The Y chromosome, because it has on it the gene for maleness, is passed on down through the male line only. Mitochondrial DNA is passed down by a mother to all her children but only daughters can pass it on: therefore it is passed down through the generations through the female line only.

SNP

Short Tandem Repeat (STR)

Male 1 :AACTGGTCATATTGACCTACTCATCATCATCATCATCATCAGGTAGTTA

Male 2 :AACTAGTCATATTGACCTACTCATCATCATCATCATCATCATCAGGTAGT

FIGURE 11.3   There are two main types of markers that are used to distinguish between Y chromosomes. These are SNPs (Single Nucleotide Polymorphisms) and STRs. A SNP is a single base change, whereas STRs are likened to a stutter in the DNA where a short motif is repeated a number to times; with the number of repeats can differ between individuals.

slowly that it is possible to draw up a large family tree of Y chromosome types based on these markers. We can use statistical methods based on these mutation rates to place the ancestor of all Y chromosome types found today to be in Africa ~100,000 years ago. Each branch of the tree is determined by a binary marker and each of these branches is further subdivided by more binary markers. Thus it is possible to 'type' a man's Y chromosome type by determining if he is ancestral (has the old form at that particular place on his Y chromosome) or derived (new form, or allele, at that marker) for a number of these differ-ent markers and determine which part of the tree his Y chromosome belongs to. The different branches of the tree are broad groups known as haplogroups.

Because of their very slow mutation rate, haplogroups only allow us to discriminate between men at a very broad level: we know for example, that many millions of men will have Y chromosome types that fall into haplogroup R1b for example. Therefore we need a way of differentiating between men at a finer scale and we do this using a type of marker known as an STR (short tandem repeat). This can be likened to a stutter in the DNA sequence where a particular short sequence of bases is repeated many times. A mutational event which gives rise to the deletion or addition of one of these repeated units, as the Y chromosome is being copied to be passed down from father to son, is fairly common and therefore it is possible to find many different alleles for a particular marker (for example, one man may have 12 repeats of a particular order of bases whereas the next man may have 11 repeats and still another may have 14 repeats). There are over a hundred STRs on the Y chromosome and usually many of these markers are typed to generate what can be likened to a Y chromosome fingerprint, known as a haplotype. These two different types of markers, because of their varying mutation rates, can be used in combination to examine events at varying time depths.

Next, Find Your Vikings

The Vikings arrived in the British Isles over one thousand years ago, settled here and, presumably, had descendants. But since that time there has been a large amount of immigration and population movement. This is certainly true of the Wirral and West Lancashire where the population growth has been far greater than that in many parts of Britain (Roberts 2002). This may have eliminated, or at the very least blurred, any traces of the Viking contribution to the population. This situation can be likened to a palimpsest: we are interested in the genetic impact of a population migration that occurred many hundreds of years ago, but since this time the genetic makeup of the population has been overwritten by more recent events such as population movements as a result of the industrial revolution and population growth. So how can we read the layers that we are interested in? How do we examine the genetic impact of the Vikings on the population of Britain?

Ancient DNA

One way would be to examine the DNA from skeletons of the period, an approach which can be likened to a 'direct' reading of the genetic makeup of the populations from the time of interest. However there are many problems associated with this method. The first of these is the paucity of skeletal material. Research is limited by the samples available which is the product of the number of burial sites found from the period of interest, the preservation of the bone/teeth, the nature of the burial (cremation, inhumation, single burials or multiple individuals in a single burial), and, as ancient DNA work is destructive, the value of those specimens to museums and whether those curating the samples are willing/able to allow ancient DNA work to be carried out on them.

A second problem is that after death DNA degrades: after some time there is a very small amount of intact DNA left to analyze, if any. The degree of degradation is affected by soil conditions and other factors such as temperature and moisture—certainly cold and dry conditions give the best chance of any DNA being recoverable and of longer fragments being available for analysis.

Steps must be taken to minimize the contamination of old samples with modern DNA. Any handling by a modern individual will have left large amounts of their own plentiful and intact DNA on the ancient specimen. This leads to the real possibility of the signal from the DNA of the specimen being overwhelmed by modern DNA in any subsequent experiments. Indeed the routine washing of archaeological specimens on site or back in the archaeological lab is thought to potentially introduce contamination. All samples must be carefully cleaned, the DNA extracted and experiments set up in a dedicated ancient DNA facility and all people who have been in contact with the specimen should have their DNA typed to allow these DNA sequences to be distinguished from those of the specimen itself. Ideally, specimens to be examined should be recently excavated under clean conditions to minimize the possibility of contamination from the outset, though clearly this is not always possible.

Recently, new technologies have been developed which make it possible to analyze smaller fragments and smaller amounts of DNA. This is being exploited in the field of ancient DNA and a number of studies of skeletons from the Viking period in Denmark are proving that we are on the cusp of being able to use ancient DNA techniques to address some of the questions surrounding the Viking migration. (Dissing, et al. 2007; Melchior, et al. 2008; Melchior, et al. 2010).

However, crucially for geneticists interested in the contribution of a particular immigrant group, such as the Vikings, to the genetic makeup of the population today, the particular individuals being examined using ancient DNA may not have contributed genetically to succeeding generations at all.

Modern DNA

The most widely used method for looking for genetic signatures of past populations is to sample from modern populations and extrapolate back to the most likely sequence of events to explain the modern genetic patterns. This method certainly has advantages. It is easy to collect samples: research groups offering genetic testing (especially where results are returned to individuals) are often overwhelmed by people willing to take part in a study. Sampling often takes the form of volunteers providing buccal (cheek) cells and can be done by sending out 'spit-kits' through the post. The DNA extracted from the samples is abundant, intact and straightforward to examine and it is possible to look at any section we choose—we are not limited by what DNA remains as is the case with ancient specimens. Due to their simple method of inheritance, it is the Y chromosome and mtDNA which are the most extensively studied part of the genome when examining past population histories using modern DNA.

This method is not without its own drawbacks, however. Modern samples represent the end result of all the events that have formed the population as it stands today. Questions arise as to what criteria to use when choosing who to sample to best represent the events we are interested in—sampling individuals with deep ancestry in the area is the ideal. Furthermore, researchers often wish to sample from rural rather than urban areas. As towns consistently attract immigrants from both the local surroundings and further afield, urban samples are more likely to include individuals whose ancestry is not local. Finally, few people have knowledge of where their ancestors were born, the limit often being at the grandparental level. Therefore individuals are often sampled based on parental or grandparental place of birth being within a region or with certain distance of a town giving a relatively shallow time-depth for known ancestry in the area. Such studies have concentrated on the proportion of Scandinavian ancestry in for example Iceland (Helgason, et al. 2001), the Faroe Islands (Als et al. 2006), Orkney and Shetland Islands (Goodacre, et al. 2005), the islands of the North Atlantic and the British Isles (Capelli, et al. 2003).

The Wirral and West Lancashire Study: Testing a New Method of Sampling

What follows is a summary of a pilot genetic study carried out in the Wirral and West Lancashire and all references to the Wirral and West Lancashire study refer to the publication by Bowden, et al. (2008). The aim of that study was to examine the genetic legacy of the Vikings in the Wirral and West Lancashire using modern samples. Thus we wanted to collect DNA from those who have the deepest ancestry in the area where Vikings settled. So how to find those individuals? Certainly the standard approach of sampling individuals with two generations of ancestry in the area would at least get around the problem of very recent immigration to the area. But to find individuals with deeper ancestry in the area a new selection approach suggested itself, the foundation of which rests on the tight link between surnames and the Y chromosome. So let's examine these two linked elements in turn.

Surnames

The practice of using hereditary surnames was brought to Britain by the Normans following the invasion of 1066. In France, family names had been in use by the aristocracy for a couple of generations. As they had in Normandy, the Norman lords here used them as a way of solidifying claims to land and ensuring inheritance and therefore the first hereditary surnames were often derived from place-names, e.g., Roger de Poitevin (McKinley 1990). Hereditary surnames in England were first adopted by the wealthy landowning families, and the use of them then gradually filtered down to the rest of population becoming commonplace in the south earlier than in the north of the country. For many years, however, people still had non-hereditary bynames and it wasn't until 1350AD that many families had surnames with it becoming rare not to have a surname by 1500AD (McKinley 1990, Redmonds 2004, Hey 2000). This gives an average age for surnames of ~700 years.

The surnames that people used came from a variety of sources broadly grouped into six categories. Many surnames derive from first names, usually from the father, with a child being known as theirs, giving us surnames such as Johnson and Roberts. Other surnames derived from a person's occupation (Smith, Thatcher), or nicknames, (Jobling, Brown) from how a person looked or behaved, from a relationship (Cousins). Others got their surnames as a way of indicating where they came from (locative surnames from a place (Raby, Lancaster) or where they lived (from topographical features of the landscape (Beck, Bush). Some surnames have multiple derivations, for example being derived from both a personal name and a place name (Redmonds 2004). Thus each surname in your family tree carries with it a story about a feature of an ancestor's life.

Two interesting characteristics of surnames are their large overall number and their individual rarity. The 1881 census listed over 400,000 surnames, a number which is thought to be even greater now, mostly as a result of immigration. Perhaps unsurprising then is the fact that most surnames are rare: the most frequent 500 surnames are carried by ~45% of the population meaning that the vast majority of surnames have only a small number of carriers. Indeed the average number of carriers for a surname is around 70 people. Finally, while some surnames (particularly the common ones) are found across the country, many surnames have a 'home' where the surname originated potentially many hundreds of years ago and around which most people with the surname still live today. Thus a person's surname often contains within it information about where his or her family can trace their ancestry back to.

The Y Chromosome

The nexus between the field of surname history and genetics arises in the form of a particular piece of our DNA: the Y chromosome The job that the Y chromosome does has an impact on how it is inherited: it can, because it carries the gene that determines maleness, only be inherited from father to son and therefore the Y chromosome that a man carries is essentially that of his father, which was that of his father and so on back through time.

Not all men have the same Y chromosome. Mutations, errors in the DNA sequence that are generated when the Y chromosome is being copied, are passed down to the next generation, introducing variation in the types of Y chromosomes seen among men. Perhaps the easiest way to think of this mutational process is to think of the Y chromosome as a story that is told by a father to his male offspring. We know the story originated in Africa over 100,000 years ago and since this time, fathers have passed down the story through the generations. Sometimes a father would change the story very slightly when telling it to one of his sons and this son in turn would tell this new version of the story to his male offspring, such that now it is possible to find different versions of the same story in different parts of the world, and with more closely related men having their own more similar versions of the story. Geneticists are able to use molecular tools to determine what a man's Y chromosome story (type) is.

One might expect then that as surnames became heritable, both a man's surname and his Y chromosome type would be passed down through the generations hand in hand such that if only one man took on particular surname many hundreds of years ago, then all men alive today with that surname should be descended from him and have his Y chromosome type (or a very similar one). If this were indeed the case, that a link exists between a surname and a particular Y chromosome type, then sampling men with old surnames from the Wirral and West Lancashire should give a better representation of the types of Y chromosomes found in those regions, ~700 years ago.

A Link between Surnames and the Y Chromosome?

At the outset it could be surmised that, at least for common surnames, it was unlikely that there would be a simple link between a surname and a single Y chromosome type. For surnames such as Smith, from the occupation of blacksmith, and with the likelihood of there being a blacksmith in most towns and villages, the surname would have been taken on a number of different times, by different men, all with their own individual Y chromosome types which they would then pass down to their male-line descendants. But what about rare names? As we know, most surnames are rare: did each one have just one Y chromosome type associated with it; that, presumably, of the original founder several hundred years ago? With an average surname-age of 700 years would we still find a link after all this time? Perhaps events such as illegitimacy (where a child would have one man's surname but another man's Y chromosome) or adoption, or inheriting the mother's surname would break the link between a surname and a Y chromosome type. Other factors will affect the types of Y chromosome associated with any one surname today. Genetic drift, the difference in the number of male offspring that men have, means that some men will have many sons, passing on their surnames and Y chromosomes many times whereas, at the other end of the spectrum, other men will have no sons to carry on their surnames and Y chromosome types: a phenomenon known as 'daughtering out' in genealogical circles. Finally, we know that tiny mutations (changes in the story) would have occurred on the Y chromosomes in a true lineage over time which would have to be taken into account. How would all these factors affect the types and number of Y chromosomes types associated with a surname?

The Surnames Studies: The 'Pairs' and 40 Surnames Studies

The first step was to do a 'pairs' study asking the question: what do we find when we take two men at random from around the country with the same surname but who don't know themselves to be related? Do they also have the same Y chromosome type with both their surname and Y chromosome type being descended from the same ancestor? To carry out the pairs study, surnames were chosen from across the frequency spectrum, from very common names (Smith) down to lower frequency names (Rivis—with only 50 people carrying this surname in Britain). 150 pairs of same-surnamed men were collected and their Y chromosomes analyzed using 17 STRs and 13 SNPs and it was found that overall, 24% of the pairs appeared to share a common ancestor through their surname. Given that we expect common names to have more than one founder (and so two men taken at random with a common name might not be expected to have inherited their surname and Y chromosome from the same ancestor) it was interesting to look at the lower half of the frequency spectrum where surnames are more likely to have had a single founder: here just under 50% of the pairs had results consistent with them sharing a common ancestor through their surname. Thus despite the length of time since surnames were established and all the factors that could affect the link between surname and Y chromosome a link existed and that link was stronger the rarer the name. An in-depth study of 40 surnames (with known spelling variants) using the same STRs and additional SNP markers then followed where many men with the same surname, but who did not know themselves to be related, were recruited. A similar pattern was seen. Among the common names, very few men appeared to be related. However, when looking at middle frequency surnames, 'descent clusters' began to appear: groups of men with the same surname (or with surname spelling variants) who had identical or near identical Y chromosome types suggesting that they all shared a common ancestor through their surname. The picture was even more striking among the rarer names. It was often the case that a single surname was dominated by a single Y chromosome type with related types. These were large groups of men who were apparently unrelated but the genetic analysis proved that they all shared a common ancestor through their surname. The best example from the set is provided by the surname Attenborough (with the spelling variant Attenborrow) where 31 men from around the country bearing either of these names, and who did not know themselves to be related, were sampled. Out of these 31, 27 had identical or near identical Y chromosome types such that they all form part of an as-yet-to be constructed family tree. Thus both studies gave the same result: a strong link exists between surname and Y chromosome type with the link being stronger the rarer the name (King et al., 2006; King and Jobling, 2009, 2009a; Redmonds et al., 2011).

Thus the link between surname and Y chromosome provides us with a method to sample men with deep ancestry in an area putatively back to the time when hereditary surnames were being adopted. So how to apply this surname sample method? It is important to point out that the Vikings did not use hereditary surnames and that the use of hereditary surnames in this country did not become widespread until hundreds of years after the Viking rule.. In Ireland, where the early period of surname adoption overlapped that of the Viking period, this provided an opportunity for sampling men carrying Irish surnames thought to have Norse roots. One small study sought to do just this, sampling men bearing surnames thought to have putative Norse origin, though the results did not suggest substantial Scandinavian ancestry among them.

The Wirral and West Lancashire Study

This study formed a pilot project first to test the validity of using surnames as a method for sampling and to carry out an in-depth study of a small area of the British Isles to test for the genetic legacy of the Vikings. For this reason two different sampling methods were used. The first of these is the one that is often used, that of two generations of residence in the area, getting round the problem of very recent immigration in that the person's ancestry is known to extend back for at least that time span. Surnames were ignored for this sample set. However, the urban and coastal nature of much of the Wirral and West Lancashire, is likely to mean that a proportion of these sorts of samples will be made up of individuals who moved to the area within the last few hundred years, not least as part of the population movements that took place during the Industrial Revolution. As such, this was unlikely to be a close representation of the population of the distant past. By simply sampling men living there today we are likely to get a very blurred picture of the genetic contribution of the Vikings.

The second sampling method used was that where people we asked if they had two generations of residence combined with having a surname which was known to have existed in the Wirral or West Lancashire long ago. Although this approach does not guarantee that sampled men are the descendants of these past surname bearers it does at least make it more likely. This is particularly true for men bearing surnames that are rare and largely confined to this area. For surnames that are common the link is more tenuous as it is much more possible that the men sampled now are not descendants of a local original surname bearer but of someone who moved into the area more recently but who also carried this common name.

Surnames and Sampling

Documentary sources were consulted in order to find lists of surnames that were present in the medieval period. For the Wirral, these sources yielded 236 names compiled from the subsidy rolls from 1542, court proceedings from 1353, and ale-house licensing records from 1572. For West Lancashire a list of 269 surnames was obtained from a document listing those resident in Bickerstaffe, Burscough with Marton, Ormskirk, Scarisbrick with Hurlton, Skelmersdale, and Westhead with Lathom, who promised to contribute to the stipend of the priest of the altar of Our Lady at Ormskirk in 1366 and supplemented with surnames deriving from old local place names or place name elements. Thus the surnames themselves are not required to be Scandinavian in origin, though many of the surnames do contain Scandinavian elements from local place names from which the surnames were derived. It was necessary to convert some of the surnames to their modern equivalents to allow selection of study participants.

There were two sample sets from each location, each collected based on the two different criteria: 'modern' samples with no account taken of surname and 'medieval' men carrying a surname known have been in the region during the medieval period based on the old surnames lists. The sample sizes were as follows: Wirral modern (100 individuals) and medieval (37 individuals) and West Lancashire modern (49) and medieval (42) with men each bearing one of the surnames below.

Wirral: Barker, Beck, Bennett, Billing, Bird, Bryde, Bushell, Colley, Corfe, Edmunds, Forshaw, Gill, Green, Harding, Hesketh, Holmes, Hough, Joynson, Kemp, Kirk, Lunt, Oxton, Raby, Rathbone, Richardson, Rimmer, Robinson, Sampson, Scarisbrick, Sherlock, Skinner, Taskar, Tellett, Tottey, Upton, Young.

West Lancashire: Alker, Balshaw, Bilsborrow, Brown, Carr, Charnock, Coly, Cook, Cooper, Corfe, Crombleholme, Fletcher, Gill, Gray, Hesketh, Holland, Holmes, Hulme, Jones(son), Leyland, Lunt, Melling, Molyneux, Otty, Pendleton, Penketh, Pennington, Prescott, Rigby, Rimmer, Risley, Roby, Scarisbrick, Sephton, Serjeant, Swarbrick, Thomason, Walsh, Webster, Westhead.

In order to avoid bias, including two or more men with the same surname or spelling variant was avoided. Some surnames were found in both the Wirral and the West Lancashire sets but these were not thought to be related. Additional samples were collected from mid-Cheshire under the two generations of residence criterion given that it is close to the Wirral and West Lancashire but with a lower proportion of Scandinavian place name elements.

Sampling, Y Chromosome Typing and Analysis Methods

Samples were collected using buccal (cheek) cell swab kits and participants were asked to fill in a genealogical questionnaire and consent form. The kits were then returned to the Department of Genetics at the University of Leicester. Here DNA was extracted from each sample and Y chromosomes typed using 13 binary markers defining 13 haplogroups, of which 9 were found in the dataset. Results of genetic studies from other parts of the country were taken from the literature to compare with those of this study. As not all labs use the same markers, some haplogroups were combined to give equivalent groupings across the datasets. In addition to these binary markers, six Y-STRs, were also typed.

The datasets from this study, both modern and medieval, were compared against each other and against data from other parts of the British Isles in terms of the levels of Norwegian contribution to each set. In the first instance, data representing the genetic makeup of the region prior to the Vikings having arrived was sought. For this set precedent (Goodacre, et al. 2005) was followed and pooled data from central Ireland and central Scotland were used for this purpose. Data from Orkney and Shetland were also included as these were known to show substantial Scandinavian admixture (Goodacre, et al. 2005). Also included were data from regions close to the Wirral and West Lancashire, mid-Cheshire, Penrith, the Isle of Man and Llangefni. Finally, a dataset representing the Norwegian Vikings was required and for this data from modern Norwegian men was used.

The Results

When examining the Y chromosome haplogroups of all the sets, it could be seen that they were typical of Western European Y chromosome types. There was one exception: there was a single individual bearing a Y chromosome type known as E1b1a1 (E-M2) in the 'medieval' set from the Wirral. Not only was his surname in the old medieval lists but he knew of at least four generations of ancestry in the area. As his Y chromosome type is one which is typical of sub-Saharan African populations it likely arrived in the region through an African migrant at some point in the past. This is not the first instance of such an occurrence. In a separate study a number of men bearing a surname which traced back to Yorkshire were found to have a rare African Y chromosome type (King et al., 2007). Both these instances could be the result of African presence in Britain through the Roman occupation, or the African slave trade. Given Liverpool's prominent involvement in the slave trade, the latter seems the most likely route.

The first question to ask was whether the sampling method led to differences in the types of Y chromosomes seen in the Wirral and West Lancashire data. Two striking differences emerged: not only were the medieval samples more diverse in terms of the types of Y chromosome types seen but they had a higher proportion of a particular Y chromosome type, haplogroup (hg) R1a1. Interestingly hgR1a1 is a Y chromosome type that is frequent in Norway (35%) but rare in most of mainland England and Wales (Capelli et al. 2003). In West Lancashire the proportion of modern samples with this Y chromosome type was 2% rising to 17% (a statistically significant increase) in the medieval group, with the values for Wirral being 4% and 14% (and increase just short of being statistically significant) respectively. These results are consistent with a higher proportion of Norwegian lineages being seen in the medieval versus the modern samples (Figure 11.4).

It is possible to use statistical methods to look at the 'genetic distances' between populations using a statistic known as F_{ST} (when looking at the haplogroup level) and R_{ST} (when using STR haplotypes) and examine these using what's known as a multi-dimensional scaling plot. In both analysis the trend is similar: the medieval samples are orientated closer to the Norwegian, Shetland and Orkney (known to have experienced substantial Scandinavian input) samples than to the modern samples (Bowden, et al., 2008).

We can also using a method known as population admixture analysis (Helgason, Hickey et al. 2001) which quantifies the genetic contribution of 'parental' populations to a 'hybrid' population. In this instance, the Wirral and West Lancashire populations can be considered to be hybrid populations with contributions from the indigenous population, and the incoming Norse Viking population. For this analysis the samples representing the two 'parental' populations were those from central Scotland and central Ireland, representing the indigenous British population and samples from modern Norwegian men representing the Viking contribution. The results were striking. The modern samples from the Wirral and West Lancashire both showed 38% Scandinavian admixture whereas the medieval samples for West Lancashire showed (51±4)% and the Wirral showed (47 ±5)%. Removing the more frequent surnames (over 20,000 carriers), known to be less reliable as surrogates for past populations,

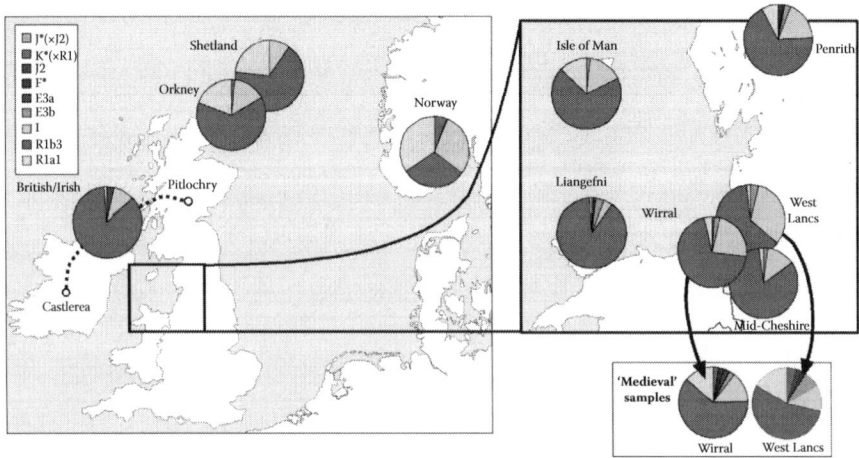

FIGURE 11.4 This shows the frequency and distribution of haplogroups in the Wirral, West Lancashire and control populations. It highlights the significant difference in the haplogroup frequencies found between the modern and medieval datasets collected for this study. From Bowden et al. (2008).

increased the proportions of Norwegian ancestry still further to (51±6)% for the Wirral and (53±5)% for West Lancashire.

Conclusions

The findings of the Wirral and West Lancashire study suggested that using this new surname-based sampling technique is providing us with a means of leap-frogging over recent population movement and increase to obtain a better picture of the genetic make up of the population in the past. Certainly the results corroborate the archaeological, linguistic and place-name evidence that this area had greater Scandinavian influence that other parts of of Britain. And the surname-based sampling, which we would expect to give a better representation of the genetic makeup of the population of several hundred years ago, shows an increase in the Scandinavian influence perhaps now diluted by more recent population movement and increase.

The Future

Higher resolution typing of Y chromosomes allows us to undertake a more detailed analysis of the genetic make-up of the population. While it isn't scientifically sound to link any one Y chromosome type to a particular historical-cultural group, statistical analysis of greater numbers of samples alongside further DNA typing should allow us to expand on this study and conduct a more comprehensive analysis of the Viking contribution to the present day population of this region. The author is currently leading a much larger genetic study examining the genetic legacy of the Vikings across the north of England.

Furthermore, with the recent development of techniques that allow for the improved analysis of ancient DNA, we now have the ability to examine the DNA from Viking age skeletal remains and directly determine the genetic makeup of these past populations. Genetic analysis could be combined with stable isotope analysis potentially allowing us to answer questions about where these individuals arrived from, where they grew up, whether the populations were made up of family groups and whether there were there fluctuations in the genetic makeup of these populations over time and, if so, why? We are on the cusp of an exciting new period in Viking Age research.

ACKNOWLEDGMENTS

The author would like to thank Professor Mark Jobling for his comments and advice on the manuscript, Dr Helen Foxhall Forbes for advice on historical references and Professor Kevin Schürer for advice on references and for discussions about surnames in general.

REFERENCES

Als, T. D., Jorgensen, T. H., Børglum, A. D., Petersen, P. A., Mors, O. and Wang, A. G. (2006). Highly discrepant proportions of female and male Scandinavian and British Isles ancestry within the isolated population of the Faroe Islands. *European Journal of Human Genetics* 14, 497–504.

Ballantyne, R.M. (1869). *Erling the Bold*, Project Gutenberg.

Barber, C., Beal, J. and Shaw, P. (2009). *The English Language*. Cambridge University Press, Cambridge, UK.

Bowden, G.R., Balaresque, P., King, T.E., Hansen, Z., Lee, A.C., Pergl-Wilson, G., Hurley, E., Roberts, S.J., Waite, P., Jesch, J., Jones, A.L., Thomas, M.G., Harding, S.E. and Jobling, M.A. (2008). Excavating past population structures by surname-based sampling: The genetic legacy of the Vikings in north-west England. *Molecular Biology and Evolution* 25, 301–309.

Capelli, C., Redhead, N., Abernethy, J.K., Gratrix, F., Wilson, J.F., Moen, T., Hervig, T., Richards, M., Stumpf, M.P.H., Underhill, P.A., Bradshaw, P., Shaha, A., Thomas, M.G., Bradman, N. and Goldstein, D.B. (2003). A Y chromosome census of the British Isles. *Current Biology* 13, 979–984.

Cavill, P., Harding, S.E. and Jesch, J. (2000). *Wirral and Its Viking Heritage*. English Place-name Society, Nottingham, UK.

Dissing, J., Binladen, J., Hansen, A., Sejrsen, B., Willerslev, E., Lynnerup, N. (2007). The last Viking King: a royal maternity case solved by ancient DNA analysis. *Forensic Science International*. 166, 21–27.

Griffiths, D. (2010). *Vikings of the Irish Sea*. The History Press, Stroud, UK.

Fellows-Jensen, G. (1985). *Scandinavian Settlement Names in the North-West*. Københavns Universitet, Copenhagen.

Fellows-Jensen, G. (1992). Scandinavian Place-names of the Irish Sea Province, In (Graham-Campbell, J., ed.), *Viking Treasure from the North-West: The Cuerdale Hoard in its Context* (National Museums and Galleries on Merseyside, Liverpool UK, 31–42.

Fellows-Jensen, G. (2000). Vikings in the British Isles; the place-name evidence. *Acta Archaeologica*, 71, 135–146.

Goodacre, S., Helgason, A., Nicholson, J., Southam, L., Ferguson, L., Hickey, E., Vega, E., Stefánsson, K., Ward, R. and Sykes, B. (2005). Genetic evidence for a family-based Scandinavian settlement of Shetland and Orkney during the Viking periods. *Heredity* 95, 129-135.

Haywood, J. (1995). *The Penguin Historical Atlas of the Vikings*. Penguin, London.

Helgason, A., Hickey, E., Goodacre, S., Bosnes, V., Stefansson, K., Ward, R., and Sykes, B. (2001). MtDNA and the islands of the North Atlantic: Estimating the proportions of Norse and Gaelic ancestry." *American Journal of Human Genetics* 68, 723–737.

Hey, D. (2000). *Family Names and Family History*. Hambledon and London, London.

Higham, N.J. (1992). Northumbria, Mercia and the Irish Sea Norse, 893–926. In (Graham Campbell, J., ed.) *Viking Treasure from the North West: Cuerdale Hoard in its Context*. Liverpool. National Museums and Galleries on Merseyside, Liverpool UK, 21–30.

King, T.E., Ballereau, S.J., Schurer, K. and Jobling, M.A. (2006). Genetic signatures of coancestry within surnames. *Curr. Biol.*, 16, 384–388.

King, T.E., Parkin, E.J., Swinfield, G., Cruciani, F., Scozzari, R., Rosa, A., Lim, S., Sue, U., Tyler-Smith, C. and Jobling, M.A. (2007). Africans in Yorkshire?—The deepest-rooting clade of the Y phylogeny within an English genealogy. *Eur. J. Human Genet.*, 15, 288–293.

King, T.E. and Jobling, M.A. (2009a). What's in a name?—Y chromosomes, surnames, and the genetic genealogy revolution. *Trends in Genetics* 25, 351–360.

King, T.E. and Jobling, M.A. (2009). Founders, drift and infidelity: The relationship between Y chromosome diversity and patrilineal surnames. *Molecular Biology and Evolution,* 26, 1093–1102.

McEvoy, B., Brady, C., Moore, L., and Bradley, D. (2006). The scale and nature of Viking settlement in Ireland from Y-chromosome admixture analysis. *European Journal of Human Genetics*, 14, 1288–1294.

McKinley, R. (1990). *A History of British Surnames*, Longman, London.

Melchior, L., Kivisild, T., Lynnerup, N. and Dissing, J. (2008). Evidence of authentic DNA from Danish Viking Age skeletons untouched by humans for 1,000 Years. *PLoS ONE* 3(5).

Melchior, L., Lynnerup, N., Siegismund, HR., Kivisild, T. and Dissing, J. (2010). Genetic diversity among ancient Nordic populations. *PLoS ONE* 5(7).

Redmonds, G. (2004). *Names and History: People, Places and Things*. Hambledon and London, London.

Redmonds, G., King, T.E. and Hey, D. (2011). *Surnames, DNA and Family History*. Oxford University Press, Oxford, UK.

Richards, J. (2004). *Viking Age England*, Tempus. Stroud, UK.

Roberts, S. (2002). *A History of Wirral*. Phillimore, Chichester, UK.

Schurer, K. (2004). Surnames and the Search for Regions. *Local Population Studies* 72, 50–76.

Smail, D.L. and Gibson, K. (2009). Sermons and learned commentary on anger and vengeance. In (D.L. Smail and K. Gibson, eds) *Vengeance in Medieval Europe: a Reader*, University of Toronto Press, Toronto, Canada, Chapter 9.

Sulem P, Gudbjartsson, D.F., Stacey, S.N., Helgason, A., Rafnar, T., Magnusson, K.P., Manolescu, A., Karason, A., Palsson, A., Thorleifsson, G., Jakobsdottir, M., Steinberg, S., Pálsson, S., Jonasson, F., Sigurgeirsson, B., Thorisdottir, K., Ragnarsson, R., Benediktsdottir, K.R., Aben, K.K., Kiemeney, L.A., Olafsson, J.H., Gulcher, J., Kong, A., Thorsteinsdottir, U. and Stefansson, K. (2007). Genetic derminants of hair, eye and skin pigmentation in Europeans. *Nature Genetics* 39, 1443–1452.

12

Figuring It Out: Further Work on the Neston Cross Fragments, Cheshire

Roger White

CONTENTS

ABSTRACT   The group of five carved 10th to 11th century cross fragments at Neston, Cheshire, constitute an unusual and interesting group of great iconographic value. In form, they belong to a well-defined type of circle-headed crosses known on the margins of the Irish Sea, but in design they are distinct in the range and number of figural carvings they bear. This chapter seeks to explore whether the author's suggestion in an earlier article of a tentative reconstruction of one cross based on two of them is accurate (White 1988). This chapter also considers the question of the significance of the overall group in terms of the iconography and what it can tell us of the society that created and used these monuments.

Introduction

In 1875, during the rebuilding of the Church of Sts. Mary and Helen, Neston, Cheshire, four fragments of carved stones were discovered and briefly published (Smith 1875, pp. 88-93). In a pioneering article on the group, J.D. Bu'Lock (1959) recognised them as examples of circle-headed crosses of Viking date (10th to 11th centuries) intended for grave markers.

Bu'Lock's original article and later work (Bailey 1980, pp. 177–180) examined the wider context of the group and established that there was a cluster of such crosses around the Irish Sea coast, from Anglesey to the Solway Firth in Cumbria with an outlier in Yorkshire and related examples in Cornwall. A fifth fragment, originally built into a window in the belfry at Neston perhaps in the 14th or 15th century, was removed in 1986 and proved to be the largest and most iconographically complex example of the group (White 1988, pp. 49–51).

The stones in the six figures described below were published recently in the *Corpus of Anglo-Saxon Sculpture* (Bailey 2010, pp. 85–90). The publication firmly placed the group in its wider context artistically and culturally, so detailed descriptions of the pieces are not required here. The following brief outline discussion relies on the numbering of the cross fragments and their carved faces established by Bailey.

Stone 1 comes from the base of a tapering shaft. Face A is carved with a bearded priest in full vestments holding up a chalice in one hand while the other presumably held the host, although the hand is missing (Figure 12.1). A maniple dangles from this wrist. On the reverse (Face C) is an irregular

FIGURE 12.1 Stone 1, Face A. (Photo courtesy of author.)

interlace based on the Stafford knot. The narrow side panels (Faces B and D) carry ring-encircled twist and a step pattern—the same patterns seen on the other principal fragments.

Stone 2 comes from the top of a tapering shaft and incorporates a segment of the circle head. Face A carries a winged figure, presumably an angel, lying horizontally and dressed in a pleated kirtle (Figure 12.2). On Face C are two figures apparently fighting. They appear to be armed with daggers and one pulls the other's hair (Figure 12.3). The narrow sides carry the same patterns as Stone 1. Face B and Face D (Figure 12.4) are described below in the Critical Analysis section.

Stone 3 is the largest and comprises the majority of a tapering shaft including its plain dressed foot. Both broad faces carry figural carving and the narrow sides are decorated in the same patterns as the other stones. Face A is dominated by a stag (Figure 12.5). At its throat is an attacking hound, while behind and above it is a figure in a tunic stretching to his knees and wielding a spear that appears to pass through the stag's body to emerge under it, the tip of the spear piercing the ground line below.

FIGURE 12.2 Stone 2, Face A. (Photo courtesy of author.)

FIGURE 12.3 Stone 2, Face C. (Photo courtesy of author.)

FIGURE 12.4 (a) Stone 2, Face D. (b) Stone 2, Face B. (Photos courtesy of author.)

FIGURE 12.5 Stone 3, Face A. (Photo courtesy of author.)

Above and to the right of this group are two figures with missing heads, on the left a female with a pigtail wearing a pleated skirt and broad triangular sleeves. She apparently embraces a man on the right wearing a triangular kirtle. On Face C are two scenes separated by a ground line into two panels (Figure 12.6). The lower one shows two horsemen, each carrying a couched lance. They are riding toward each other with their lance heads crossing. The men have round heads and long tunics, perhaps representing chain mail shirts. The upper panel has two quadrupeds, one chasing the other which is looking back at it.

Stone 4 represents the base of another tapering shaft including the substantial dressed foot of the stone. The carved panel on Face A consists of a six-strand plait while Face C shows fragmentary remains of an unidentified knotwork. Face D is lost while Face B shows the normal step pattern.

Stone 5 is the only fragment derived from a cross head and its interpretation relies on better preserved examples from others in the Cheshire group, such as a cross head from Hilbre Island (Bailey 2010, p. 81, illustrations 172–175) or the group at St John's Chester (ibid. pp. 63–64; illustrations 81–84). The stone has been dressed to form a rectangular block, clearly of a size and shape to be used as an ashlar. Face A has a triquetra knot and possibly traces of a second, a design repeated on Face C. Face B indicates traces of a step pattern while Face D has a two-strand interlace. These designs are probably the same as the border decorations in the rest of the group, as is the cable border seen on all the edges of the carved fragments.

FIGURE 12.6   Stone 3, Face C. (Photo courtesy of author.)

Critical Analysis of Tentative Reconstruction

An initial study of the group in 1986 suggested that the new fragment removed from the belfry was the lower part of Stone 2. The match was suggested by the fact that the two fragments appeared to share the same taper and shaft thickness as well as the same type of decoration on Faces B and D. There were apparent iconographic links too, with hunting or fighting apparently shown in three registers on Face C, while Face A appeared to show a stag hunt, an embracing couple, and an angel (White 1988, plates 1 and 2).

At the time, detailed study of the pieces was not easily achieved and the hypothesis remained untested. However, it has become apparent from further work by Professor Richard Bailey and independent work by the author that this reconstruction is impossible.

The key to this conclusion was a detailed analysis of the border decoration rather than the main faces of the slabs. This detailed examination was made possible by a visit to the stones to take measurements in situ and by making use of three-dimensional laser scans of the stone obtained by the Conservation Technologies unit of the National Conservation Centre in 2007 (White 2013).

Comparison of the stones demonstrated significant differences between the slabs despite the identical patterns. On Stone 3, Face D, seven elements of the step pattern can be counted. Measured from the bottom of the stone, each element is 87, 88, 88, 78, 107, 104, and 95 mm. The median value of these is 93.5 mm. Face D on Stone 2, however, has only two surviving units of the step pattern but these are carved in a significantly different fashion. The short upright is slightly slanted rather than upright as on Stone 3 of Face D, rendering the elements more stretched in comparison to the other stone: 151 and 145 mm (Figure 12.4a).

This stylistic difference alone is enough to rule out these two patterns being from the same sculpture, as it is improbable that the carver would have so radically changed the style of the step pattern in the relatively small interval represented by the missing part of the stone. On Face B, further stylistic differences can be seen. The pattern here is more complex to measure, consisting of a ring encircled twist with a lengthening ('glide') between the rings. On Stone 3, the full length of one element of the pattern is 240 mm. The three surviving knots are 103.4, 92, and 93 mm and the measurement from the base of the complete pattern to the top of the stone is 144 mm.

Not enough of the pattern survives on Stone 2, Face B to enable a complete comparison to be made, but the surviving knot is smaller—88mm (Figure 12.4b). This discrepancy may be explained by the tapering space available at the top of the shaft. However, from the regularity of the patterning, it can be estimated that the missing element of the ring must be around 42 mm in the case of Stone 2 or 192 mm on Stone 3. However, when one aligns the two stone fragments so that their edges line up with each other (which can be done accurately through the laser scans), the amount of stone missing can be calculated as 85 mm on Face B and 60 mm on Face C.

It should also be noted that the close examination of the stones and scans clearly demonstrated that the cable borders of the stones also differed in the angles of cutting, again demonstrating that they could not be from the same monument. One final measurement was discovered. Stone 2 preserves an arc of the circle head and from this arc it was possible to calculate its diameter as 231 mm. This dimension was found to exactly match that of Chester St. John's Cross 2 (Bailey 2010, illustrations 81–84).

Neston Group in Its Regional Context

While the general form of the circle-headed cross can be seen to extend geographically around the Irish Sea littoral, from Anglesey via north-west England to the Solway Firth, regional groupings have been recognised since Bu'Lock's work (1959). These include Cheshire and Cumbria groups, the former having additional pellets in the spandrels of the cross while the latter always have pierced spandrels. The circles are also decorated differently: pellets and step patterns in the Cheshire group and plait work in the Cumbria group, but they share in common the projecting square bosses outside the ring of the circle that signify the projecting arms of the cross within.

The outliers in Yorkshire and in Wales show combinations of these traits demonstrating that both sub-groups were likely to have been carved in roughly the same period (Bailey 1980, pp. 177–181). In his more recent discussion of the Cheshire Group, Bailey adds another example related to this group from Walton on the Hill (formerly Lancashire, now Merseyside) and also includes the three related examples from Flintshire which in the immediate pre-conquest period was administratively part of Cheshire (Bailey 2010, pp. 31–33). These examples from further afield, however, differ in detail from the main group confined to the Wirral and Chester.

Also added to Bailey's discussion are examples of similar crosses at Maughold on the Isle of Man and other related crosses at various sites in Ireland; these are linked more to the Cumbrian rather than the Cheshire sub-group (ibid., pp. 32–33). Overall, these new discoveries and parallels only further add to the impression of the group as one strongly focused on Viking-period connections across the Irish Sea, with the Yorkshire outlier providing evidence of the important link with the Viking community based in York.

It is possible to speculate that the Chester and Wirral groups originated in a sculptural team based in Chester, and perhaps at St. John's where the largest group of stones is to be found. In favour of St. John's as a centre for carving is the proximity of a quarry and the presence of an apparently unfinished cross, St. John 6 (Bailey 2010, pp. 33 and 67). Links between the St. John's group and those in Neston can be seen in the interlace patterns on St. John Stone 3C and Neston Stone 4C, and in the demonstration that the circle heads are of the same diameter at both sites, suggesting layouts of the designs in the same fashion using the same tools.

Finally, the identification of at least one figural group at St John's, Fragment 8, even if the subject is in dispute, provides another important link (Bailey 2010, p. 68). One major difference can be seen in the relative thicknesses of the circle-headed crosses: those at St John's are far more substantial than those at Neston but this presumably relates to differences in the bedding layers within each quarry from which the stones were taken. The locations of these are unknown, but both are now identified as from the Bunter (Chester Pebble Beds) sandstone formation (Bristow 2010, pp. 13–17, Figure 5).

Interpreting the Neston Stones

The realisation that the proposed Neston Cross reconstruction is not possible now restores the surviving group as comprising two non-figural fragments from a head and shaft and three figured shaft fragments giving a total of four or five crosses (the head fragment could conceivably belong to one of the other four crosses represented so there is still a minimum of four crosses represented). The three figured crosses form the focus of discussion here with a view to determining their subject matter and hence suggesting their potential purpose within the community.

Interpretation of the three fragments is unclear, with perhaps one exception: Stone 1. This clearly shows the figure of a priest in full vestments and, importantly, holding aloft a chalice. The other upraised hand is missing but from the wrist dangles a maniple, strongly suggesting that the other hand held a consecration wafer. The scene depicted is thus clearly the high point of the mass: the elevation of the consecrated host when the priest presents the body and blood of Christ to the assembled congregation immediately before Communion is taken. Whilst there can be no doubt of the religious significance of the symbolism, why is it depicted in this way on a cross shaft?

These stones were always assumed to function as grave markers rather than preaching crosses although this is impossible to prove since none was found in situ. If this is the case, then we may be looking at nothing more complicated than the commemoration of the role of a buried priest rather than the presentation of a didactic message about the role and importance of the mass to the community.

Two other two stones carry much less obvious Christian symbolism despite the depiction on Stone 2 of (apparently) an angel flying horizontally across the stone's face. The significance of this figure is difficult to determine due to the loss of the rest of the stone but perhaps the angel hovered protectively over figures depicted below, although other possibilities can be suggested (Bailey 2010, pp. 86–87). J. Romilly Allen's suggestion that this fragment is part of the same monument as Stone 1 has been ruled out on geological grounds (Bailey 2010, p. 85) but can also be deemed improbable since it would be unlikely to combine the interlace pattern on Face C of Stone 1 with the figural scene on the same face of Stone 2 that depicts two men fighting.

As noted in the author's earlier discussion of these stones (White 1988, p. 52), the subjects are often said to be Jacob and the Angel although the iconographic parallels are weak and ignore the clear depiction of knives in the assailant's hands which mean that wrestling is unlikely to be portrayed. In any case, if that scene were intended, one of the figures should have had wings.

Other possibilities have been suggested by Bailey albeit without certainty (2010, pp. 87–88), including depictions of the fights of Cain and Abel and/or David and Goliath. Against the former is the fact that both men are armed; against the latter is the fact that both figures are the same size, which undermines but does not rule out these identifications.

As Bailey notes, it is not impossible that a secular interpretation is to be preferred here. For the moment, it is worth noting that both figures are of the same size, are clothed, and carry triangular blades, presumably knives. Physical contact is also evident: they grab each other's hair as one treads on the other's foot. This is clearly a violent fight, perhaps mortal combat, and may simply be a depiction of a battle scene (not necessarily biblical) or some kind of martial contest.

The more iconographically complex Stone 3 offers greater hopes for understanding the purpose of these stones since it is nearly complete. Indeed, it could be argued that all the pictorial elements of one side, Face A, are complete although even this is impossible to prove. This face depicts two figures, clearly a man and woman from their clothes, standing with their arms embracing each other.

Their closeness rules out the usual interpretation of a male and female confronted figure: a woman offering a drinking horn to a deceased man as seen beneath the crucifixion on the Gosforth cross (Bailey and Cramp 1988, illustration 304; Bailey 1980, pp. 130–131). The natural interpretation suggested by the encircling arms is an embrace, with the implication perhaps of a depiction of a couple; the cross from St. Mary Bishophill Junior, York shows a similar, if earlier, image (Lang 1991, pp. 83–84, illustration 216).

The restoration of the Neston cross for the 2009 exhibition (see below) demonstrates clearly that there would only be enough space on this face if reconstructed to incorporate the couple's heads, perhaps leaving the field behind and above the woman clear (Figure 12.7). This small area may have been dressed flat to carry a painted inscription to the deceased.

The remainder of the face below the couple is filled with a depiction of what appears to be the final stage of a stag hunt. The stag has been brought to bay by a hound that grips it by the throat. Above and behind, a standing man thrusts a spear that seems to pass through the animal. The spearhead appears

FIGURE 12.7 Reconstruction of Neston Stone 3, Face A based on digital scan incorporating outline of Stone 2 and modified head of Chester St. John's Stone 2.

under the stag's belly, piercing the ground line. While hunting was the pastime *par excellence* of the élite at the time, Christian interpretations of the pursuit can also be used to demonstrate that these images have a spiritual purpose (Bailey 2010, p. 89).

The upper panel on Face 3 shows one animal chasing another—possibly another hunt scene. The lower scene on Face 3 is much more unusual, depicting two mounted horsemen apparently conducting a joust. While mounted cavalrymen are known from other monuments of the period, as was discussed in my original article (White 1988, pp. 53–55), these depictions of riders are not close parallels. This scene appears to match in all respects the idea of jousting, and the only missing elements are shields. Since the 1988 article was written, a biblical parallel has been suggested to the author by Professor Jane Geddes based on an illustration in the St. Alban's Psalter (Geddes 2013).

Folio 72 depicts two armed horsemen riding toward each other. Their lances have been broken from their first encounter and they are drawing swords to re-enter the fray. This is clearly not quite the same image but the text refers to the *holy war on earth in the church* carried out by *holy figures, armed in a manly spirit, are made friends of Christ and heavenly athletes* (Folio 71), an interpretation that Bailey elaborates and supports (2010, p. 89). While there is clearly space for a third scene at the top of the shaft, we have no clue as to what this might have been; it is idle to speculate so the space was left blank on the reconstruction (Figure 12.8).

While each individual element of Stone 3 can be interpreted in isolation for its significance, it is important not to forget that the monument was an integral whole and intended to be interpreted collectively rather than piecemeal. Similarly, it is also the case that these monuments may to a degree cross-refer to each other or at least celebrate parallel themes since they were all created at around the same time by the same artist(s) for patrons living in the same community.

In seeking parallels for Stone 3, we can turn to the wider Irish Sea context and specifically to Sandulf's Cross at Kirk Andreas, Isle of Man (Kermode 1907, pp. 194–195). Whilst not a close parallel stylistically, this complete cross exhibits similarities to the Neston example: a stag hunt with a hound jumping

FIGURE 12.8 Reconstruction of Neston Stone 3, Face C based on digital scan incorporating outline of Stone 2 and modified head of Chester St. John's Stone 2.

on the animal's back rather than at its throat, while a horseman rides in pursuit. On the other face is another mounted figure, presumably a depiction of the deceased whose name is given as Arinbjörg in the accompanying inscription. Individual domestic and wild animals are depicted flanking the central cross.

An even closer parallel is offered by the Hilton of Cadboll stone, in Easter Ross, Scotland (James et al. 2008). Its main surviving face depicts a woman riding side-saddle above a stag hunt. The stag is chased by hounds, one at its hindquarters and another at its flank, tearing at its shoulder. A spear passes diagonally through the stag's back while two armed horsemen ride above and behind the scene. Two trumpeters standing behind the riding woman complete the panel.

In discussing this scene, Isabel Henderson notes the numerous parallels for representations of a stag hunt with hounds and mounted riders including scenes at Burghead, Aberlemno, Kirriemuir, and Meigle (2008, pp. 178–189). Her lengthy consideration of the iconography of the hunting panel in the light of the whole monument specifically rules out interpretation in an entirely secular light as a *Pictish woman enjoying the hunt.* Rather, Henderson concludes with the suggestion that the prominent woman is either a depiction of Mary or a *metaphor for the virtuous life* (ibid., p. 189).

Neston Stone 3 may thus represent the embracing couple as a metaphor for the virtues of married life within the body of the Christian church. Equally it may be simply a representation of the patrons of the monument, which would fit a general trend in the Anglo-Scandinavian era for secular portraiture (Lang 1991, p. 84). Neither interpretation is mutually exclusive, of course, and thus the stone may depict a layered message conveyed to those who knew the couple commemorated in stone (if that is indeed what is depicted) as well as a metaphor for Christian marriage.

Both the Pictish and Manx parallels demonstrate clearly that the sculptors of these monuments worked from a common source of symbolism circulating within the region and thus Neston must be considered to be within this context. The decoration on Face C of Neston stone 3 can thus be viewed in same manner.

The scenes presumably allude to Christian themes but do so in a way that can be interpreted at a number of levels. They represent the triumph of Christianity and the hope that it brings to the believer but do so in ways that directly identify with motives that celebrate the lives and achievements of those they commemorate. This interpretation would work too for the stone representing the priest. The work graphically demonstrates the potency of the church and also celebrates the key role and significance of the parish priest in the newly Christianised settlement.

Other elements such as the knife fight are less clearly indicative of Christian symbolism but presumably must be interpreted in the same manner: a contemporary representation of the struggle of the believer against evil and sin. Equally, however, it is likely that at another level these scenes reflected the contemporary lives of the people they commemorate. The stones are, after all, personal monuments whose imageries were selected presumably to reflect the lives and beliefs of the deceased.

Thus the scenes of conflict could be reflections of the endemic warfare of the period that allowed the participants to establish their importance and roles in secular society. These people may have been pious Christians, but they were also warriors and the assertion of their martial prowess on their grave monuments alongside other élite activities such as hunting is an equally valid interpretation of these scenes, reinforcing their status and roles in a wider society.

Conclusions

The collection of stones at Neston represents a snapshot of a community in transition, a recently Christianised community that felt it important to celebrate its new-found faith in a manner that drew upon the wider cultural context in which the inhabitants existed. The variety of imagery is startling and indicates a strong connection to the visual and cultural repertoire of the period whose sources presumably lay in portable artefacts such as psalters whose imagery perhaps inspired the extraordinary scenes on these and other stones (Henderson 2008, p. 183).

Yet on another level, the sources of the images were also the lives of those commemorated, a local élite whose war and hunting pursuits were common across all societies in Europe at the time. The associated imagery was thus readily understood by all community members, whether secular or religious. Analysis demonstrates that the messages conveyed by the stones are not unidimensional but instead convey layered messages signifying both piety within the context of the religious sensibilities of the time and membership of an élite pursuing the normal activities of their station common across western Europe at that time.

Over time, the community's view of these monuments must have changed because at some point they were broken up and recycled into the building. We shall never know when this occurred but perhaps at some point these monuments may have been considered sacrilegious and thus removed from the sight of society. When rediscovered in the Victorian era they were brought once more into the community's consciousness, but were still perhaps irrelevant to it—museum pieces rather than monuments.

This is reflected to a degree in how I found the stones when I first saw them in 1987: abandoned in a heap by the vestry door. My invitation to see them and to write an interpretation of the group so they could be redisplayed was very much the realisation on the part of congregation that the heap of stones was not an adequate preservation of what were historic monuments.

The resulting research carried out on the fragments could also be seen to be peripheral to the community. They had precious little involvement in the interpretation and academic study of the pieces, although no doubt they appreciated the resulting display within the church. For the wider community beyond the congregation but living within Neston and beyond, the work became relevant only when it became part of Elizabeth Royles' exhibition on Vikings in the Wirral.

Working with schools to reimagine the imagery of the stones as they might have appeared (Figure 12.9) inspired a greater awareness in the children of Neston of both the monuments and those who created them. The creation of the resin replica and its placement in the open where all can see and enjoy it place these monuments once again at the heart of the community and, like the children's artwork inspired by the form of the cross, reminds us that these monuments had real meaning to society when they were carved even if all the nuances cannot now be recaptured.

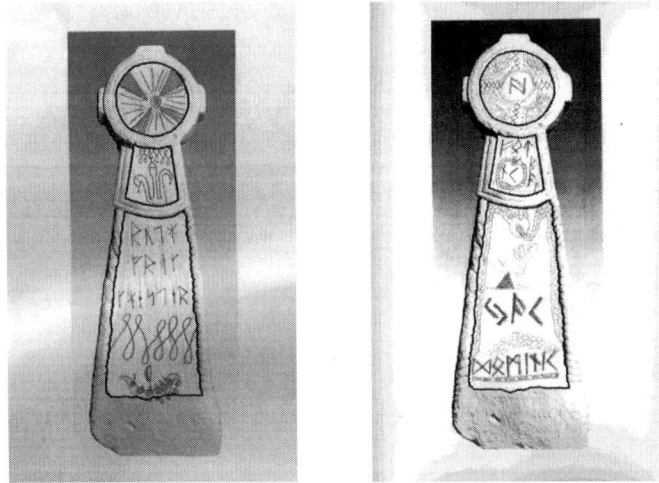

FIGURE 12.9   Viking crosses as reimagined by Neston children working with Elizabeth Royles.

REFERENCES

Bailey, R.N. (1980). *Viking Age Sculpture.* Collins, London.

Bailey, R.N., Ed. (2010). *Corpus of Anglo-Saxon Sculpture Volume IX: Cheshire and Lancashire.* Oxford, Oxford University Press.

Bristow, C.R. (2010). Regional Geology. In *Corpus of Anglo-Saxon Sculpture Volume IX: Cheshire and Lancashire*, R.N. Bailey, Ed. Oxford, Oxford University Press, pp. 10–17.

Bu'lock, J.D. (1958). Pre-Norman Crosses of West Cheshire and the Norse Settlements around the Irish Sea. *Transactions of the Lancashire and Cheshire Antiquarian Society*, 68, 1–11.

Geddes, J. Translation of St. Albans Psalter. http://www.abdn.ac.uk/stalbanspsalter/english/translation/trans072.shtml

Henderson, I. (2008). The Art-Historical Context of the Hilton of Cadboll Cross Slab. In *A Fragmented Masterpiece: Recovering the Biography of the Hilton of Cadboll Pictish Cross-Slab,* H.F. James et al., Eds. Edinburgh, Society of Antiquaries of Scotland, pp. 127–204.

Kermode, P.M.C. (1907). *Manx Crosses.* (Facsimile reprint, Pinkfoot Press, Dundee, 1994).

Lang, J., Ed. (1991). *Corpus of Anglo-Saxon Sculpture Volume III: York and Eastern Yorkshire.* Oxford, Oxford University Press.

Smith, H.E. (1875). Archaeology in the Mersey District, and Liverpool Notabilia in 1874. *Transactions of the Lancashire and Cheshire Antiquarian Society Series III,* 85–108.

White, R.H. (1988). Viking Period Sculpture at Neston, Cheshire. *Journal of the Chester Archaeological Society,* 69, 45–58

White, R.H., (2013). Resolving the Carving: Application of Laser Scanning in Reconstructing a Viking Cross from Neston, Cheshire. In *Visual Heritage in the Digital Age*, E. Ch'ng, V.L. Gaffney, and H. Chapman, Eds. London, Springer, pp. 33–41.

Index